Mind Force

On Human Attractions

STUDIES OF NONLINEAR PHENOMENA IN LIFE SCIENCE

Editor-in-Charge: Bruce J. West

Vol. 1 Fractal Physiology and Chaos in Medicine
by B J West

Vol. 2 Patterns, Information and Chaos in Neuronal Systems
edited by B J West

Vol. 3 The Lure of Modern Science — Fractal Thinking
by B J West & B Deering

Vol. 4 Physical Theory in Biology — Foundations and Explorations
edited by C J Lumsden, W A Brandts & L E H Trainor

Vol. 5 Nonlinear Dynamics in Human Behavior
edited by W Sulis & A Combs

Vol. 6 The Complex Matters of the Mind
edited by F Orsucci

Vol. 7 Physiology, Promiscuity, and Prophecy at the Millennium: A Tale of Tails
by B J West

Vol. 8 Dynamics, Synergetics, Autonomous Agents: Nonlinear Systems Approaches to Cognitive Psychology and Cognitive Science
edited by W Tschacher & J-P Dauwalder

Vol. 9 Changing Mind: Transitions in Natural and Artificial Environments
by F F Orsucci

Vol. 10 The Dynamical Systems Approach to Cognition: Concepts and Empirical Paradigms based on Self-Organization, Embodiment, and Coordination Dynamics
edited by W Tschacher & J-P Dauwalder

Vol. 11 Where Medicine Went Wrong: Rediscovering the Path to Complexity
by B J West

Studies of Nonlinear Phenomena in Life Science – Vol. 12

Mind Force

On Human Attractions

Franco Orsucci

University College London, UK
University of Siena & Institute for Complexity Studies, Italy

World Scientific

NEW JERSEY • LONDON • SINGAPORE • BEIJING • SHANGHAI • HONG KONG • TAIPEI • CHENNAI

Published by

World Scientific Publishing Co. Pte. Ltd.
5 Toh Tuck Link, Singapore 596224
USA office: 27 Warren Street, Suite 401-402, Hackensack, NJ 07601
UK office: 57 Shelton Street, Covent Garden, London WC2H 9HE

British Library Cataloguing-in-Publication Data
A catalogue record for this book is available from the British Library.

Studies of Nonlinear Phenomena in Life Science — Vol. 12
MIND FORCE
On Human Attractions

ISBN-13 978-981-277-121-6
ISBN-10 981-277-121-2

Printed in Singapore by World Scientific Printers

Contents

Preface

We shall not cease from exploration
And the end of all our exploring
Will be to arrive where we started
And know the place for the first time.

T.S. Eliot

The primary motivation for writing this book came from the need to define a unified framework to comprehend in a coherent picture, the long enterprise for understanding how the mind functions in scientific terms. We consider, and we will explain why, all the necessary empirical and theoretical tools are finally at hand. It is time then to deal with this challenge in simple and practical terms, out of the mist of preconceptions.

The first methodological and epistemological step is to consider this challenge in *complexity* terms. Only the synergy of different approaches can ensure that we might reach a possible result in a multidisciplinary project. It the past centuries, *Natural Philosophy* has tried a similar approach, but the empirical and analytical tools at its disposal were insufficient. Results were poor in science, though sometimes interesting in metaphorical terms. It is not casual that new attempts have been made with strong contributions of physicists and mathematicians, who realized that a solution might be near.

After complexity science, the technical instrument we need is the *synchronization theory*. This discipline is not new, as its roots date back to the XVII century. Yet, it is an area that has seen recent growth in its applications on chaotic and many body systems. It is exactly what we need in order to make our quest practically viable. It also gives us a way to root it into empirical situations and to build reliable models.

As we are dealing with many bodies and many variables, a way to cluster them into *fields* is another necessary step in order to recognize lines of order out of large arrays of complex interactions.

This part of the project is also facilitated by recent developments in *network theory* which defined the different kinds of organizations that complex nets can reach. Small world and free scale networks are some of the most recent definitions. These discoveries allow embracing scales of organization which can span from the molecular to the highest cognitive structures. The resulting organization will be what we call a *heterogeneous hyperstructure*, produced by the mass effect of billions of different interactions. Mind Force is the result of the causal power of collective phenomena and patterns.

This is the dynamical hyperstructure we have decided to call *Mind Force*. Though, we created this definition rather independently, we soon discovered that there were other important, though germinal, approaches following this same direction. The only thing we could think of after uncovering other contributions on *Mind Force* was to repeat the seminal motto by Bernard of Chartres, famous for being cited by Newton: standing on the shoulders of giants.

We are uncertain about the results of such a big challenge, but we do hope to have at least been able to propose further progress in the direction of Mind Force understanding. We are sending this book to print in the belief that the reader might find something worthy in the reading that will bring Mind Force from the realm of Holy Grail knowledge to the terrain of theory, and applications in psychopathology and wellbeing.

Franco Orsucci

Acknowledgments

This book was gradually raised over the last couple of years and received a final growth spurt during a conference on the same topic held in September 2008 at the Pontignano Charterhouse, University of Siena, in Italy. During this conference, we also produced the manifesto which is included in one of the appendixes of this book.

I am indebted to many people. Most of all I would like to thank Nicole Arriaga, who helped to prepare the final manuscript: I could not have finished without her outstanding contribution. I also thank the excellent editor at World Scientific Publishing, Lakshmi Narayanan, with whom I enjoy a long standing collaboration. I thank the CEO of WSP and Imperial College Press, K. K. Phua, for his vision on innovative science.

I also appreciate the work and friendship of people with whom I have had the honor of collaborating and whose enthusiasm is woven into the ideas expressed here: Walter J. Freeman, Lamberto Maffei, Peter Fonagy, Mario Reda, Joe Zbilut, Alessandro Giuliani, Fred Abraham, Sergio Rinaldi, Daniel Amit, Tito Arecchi, Harald Atmanspacher, Alfredo Ancora, Dick Byrd, Mario Fulcheri, Steve Guastello, Jeff Goldstein, Louis Pecora, Giuseppe Vitiello, Carla Fioravanti, Giorgio Parisi, Guelfo Margherita, Robert Marwan, Tullio Minelli, Chiara Mocenni, Alessandra Orsi, Luciano Pietronero, Bob Porter, Paola Redditi, Chuck Webber, Franco Scalzone, Steven Suomi, Andrea Seganti, Alan Stein, Antonio Vicino, Eveline Taylor, Angelo Vulpiani, Gemma Zontini.

Chapter 1

Time

1.1 Sympathy for Clocks

Christiaan Huygens wrote about his first discovery of synchronization in a letter to his father Constantyn, dating back to 1665:

"While I was forced to stay in bed for a few days and made observations on my two clocks of the new workshop, I noticed a wonderful effect that nobody could have thought before. The two clocks, while hanging [on the wall] side by side with a distance of one or two feet between kept in paces relative to each other with a precision so high that the two pendulums always swung together, and never varied. While I admired it for some time, I finally found that this happened due to a sort of *sympathy*: when I made the pendulums swing at different paces, I found that half an hour later they always returned to synchronism and kept it constantly afterwards, as long as I let them go. Then I put them further away from one another hanging one on one side of the room and the other one fifteen feet away. I saw that after one day, there was a difference of five seconds between them, and consequently their earlier agreement was only due to some sympathy that, in my opinion, cannot be caused by anything other than the imperceptible stirring of the air due to the motion of the pendulums. [...] and the vibrations of the pendulums when they have reached synchronism are not such that one pendulum is parallel to the other, but on the contrary, they approach and recede by opposite motion."

1.2 Music is Time

Steve Strogatz, in his brilliant book on synchronization, noted: "At the heart of the universe is a steady, insistent beat: the sound of cycles in sync. It pervades nature at every scale, from the nucleus to the cosmos"

(Strogatz, 2003). Even our bodies are participating at every scale to these rhythmic symphonies (Zerubavel, 1981).

In his recent book on *musicophilia,* Oliver Sacks stressed how our nervous system is exquisitely tuned for music. But, how much of this is due to the intrinsic physical characteristics of music itself and its complex sonic patterns woven in time? Its logic, momentum, sequences, rhythms, repetitions and the mysterious way in which it embodies emotion, all play an important role. Just how much depends on special mind/brain resonances, synchronizations, oscillations, mutual excitations, and/or feedback in the immensely complex, multi-level, neural circuitry that underlies musical perception and replay, is an intriguing matter for research (Sacks, 2007).

E. T. Hall (1983) has been an important figure in noticing that humans in all cultures are engaged in rhythmic dance. He documented the variety of rhythms by studying films of people interacting in a wide range of different situations, from laboratory to everyday life (Hall, 1983).

The role of music in the social technology of bonding is well known, having been explored by anthropologists in studies of tribal rites of passage, ordeals, and ceremonies, invariably accompanied by the use of music, drumming, dance, and other forms of predictable repetitive actions. Robert Jourdain (1997) explored the various possibilities in which music exploits our brain and body rhythms. He highlighted two different notions of rhythm: *meter* consists of a regular pattern of beats; while *phrasing* includes organic and seemingly irregular, but structured organizations of musical shapes. These different kinds of rhythm are usually superimposed but there may be prevalence to it (Jourdain, 1997).

In this complexity of biological and psychosocial interactions, we have a goal to understand how order arises. Order in space seems often to present architectures you can see but explaining order in time has proved to be more problematic and remarkably subtle. We call this kind of order synchrony. Until a few years ago, biologists, physicists, mathematicians, astronomers, sociologists and psychologists all developed the study of synchrony along parallel lines of inquiry. Synchrony, also known as sync, is an attempt to synthesize and integrate a vast body of knowledge on time orders produced in several different disciplines: a science of synchrony has been coalescing as the study of coupled oscillators. Sync is strange and beautiful at the same time. It is strange because sometimes it seems to defy the laws of physics though it relies on them simultaneously. It is beautiful because it results in a sort of cosmic dance, or rather, a universal orchestra.

Synchrony, after studies on very simple oscillators, like pendulums, tides or transistors, is now dealing with the challenge presented by nonlinear dynamics arising in systems with many variables and a large number of oscillators. Even using super computers, the collective behavior of giant systems involving many oscillators of the same type (for example neurons) or of different type (for example speech, movement and the brain) is a *terra incognita* worth a thorough exploration.

1.3 Concert for Hormones

There are invisible and inaudible concerts. Because of different kinds of forces, the tendency to synchronize is one of the most pervasive causes of order in the universe, from atoms, to animals, from people to planets. An example of how synchrony might involve different intermingled domains is provided by biorhythms. We might refer to the frequency among female friends or coworkers who spend a great deal of time together that tend to synchronize their menstrual cycle. In another book (Orsucci, 2002b), we mentioned an interesting experience on this matter, reported by one of our students. She went to stay in an apartment where two other female students already lived. They were single, friends and already synchronized in their menstrual cycle. They remained synchronized after her arrival but she didn't join them. She explained that she was emotionally distant from their close friendship, probably because she had a boyfriend, while they were single. She also didn't share a lot of time with them. Some months later, she and her boyfriend spent a great deal of time apart due to a new job he had started: she then started feeling alone and became emotionally closer to her apartment mates. They began sharing many activities, spending more time together and with time grew to be close friends. At this point, she was surprised that her menstruation cycle had become synchronized with theirs. The rhythm of periods for these three female apartment mates remained steady and in sync for months until the girl's boyfriend came back into town: then, suddenly, she became desynchronized. In this example there is a mix of different factors which could have influenced sync and de-sync: spatial proximity, living together sharing a similar schedule and a variation of emotional proximity.

A slightly different story has been reported by another woman, living very far from her sister with whom she still has a strong emotional bond. She and her sister live in two different continents. When they were living together, in their parent's house, they used to synchronize their periods, sometimes even sharing the same rhythm with their mother. Now that she lives so far away, she keeps in touch with her sister via video-chat over the Internet, and she has been noticing that when their video calls become more frequent, their periods become synchronized again. This doesn't happen when video calls are rare.

In this case, sync seems to happen due to different factors: there is no sharing space, no physical proximity, but there is a form of communication which is still carrying visual and voice signals of emotional expression.

The menstrual cycle, in biophysical terms, can be regarded as a noisy coordinated oscillation of several different hormones. The regulation of these hormones occurs via positive and negative feedback loops, as the presence of feedback is a typical feature of self-sustained oscillators. The period of oscillation, as we know it, occurs about every 28 days but can fluctuate considerably. Martha K. McClintock and her research group performed several studies which seem to support the idea of a hidden chemical communication between women (Stern & McClintock, 1998; Jacob & McClintock, 2000; McClintock et al., 2005). For example, she took swabs from the armpits of women at different points of their menstrual cycle and dabbed them on the upper lips of other women. Swabs taken from women at the beginning of their cycles tended to shorten the cycles of women who received them. In contrast, swabs taken from women at the ovulation phase prolonged the cycles of the receivers. This study might support the idea of a 'silent conversation' mediated by chemical messages, probably even pheromones.

McClintock performed another study of the menstrual cycle of 135 females from ages 17 to 25, all of whom resided at the same dormitory in a female college. The findings showed an increase of synchrony during the academic year. This confirmed that social grouping influences the balance of the endocrine system. Particularly, sub-groups of close friends defined by a self-rating scale, showed a reduction of dispersion of their frequencies. Research also showed that physiological, emotional and psychosocial factors can influence hormonal cycles. Whether this might be due to pheromones, and/or psychosocial interactions is still a matter of debate.

All mammals show forms of hormone synchronization and this is usually explained with an attempt to ovulate and conceive in step, in order to share rearing, feeding and the protection of offspring. Reproductive sync has benefits for all, if the group is cooperative. The role of pheromones in this sync process is clear for most mammals as the function of the vomeronasal organ (the part of the inner nose receptive to these chemical signals of pheromones) and its pathways to the hypophysis are proven. This is controversial for humans were this 'nose organ' is considered atrophic, and other psychosocial factors seem to influence menstrual synchronization (Orsucci, 2000).

Self-organization has been called to explain these phenomena. We know that classical science has been following, since the scientific revolution of the 17th century, the adagio "*simplex sigillum veri*": simplicity is the seal of truth. Albert Einstein posed a *caveat* using to say that "everything should be made as simple as possible, but not simpler".

1.4 Resonances

It is puzzling that in mind science statistics are still more important than dynamics, but this might be due to the difficult management of complex data and variables produced by the bio-psychosocial mind *milieu*. Psychophysical variability, for example, is usually treated as an error rather than temporal distributed information about cognitive processes. For example, seemingly random fluctuations can convey valuable information on human behavior if considered from the dynamic viewpoint.

In a beautiful book, Scott Kelso (1995) described brain activity as being based on rhythms in a "pattern forming, self-organized, dynamical system poised on the brink of instability". For example, he presented a clever finger-twiddling experiment, in which he explained using his famous model of coordination developed in collaboration with Hermann Haken and W. Bunz: the Haken-Kelso-Bunz (HKB) Model. The HKB model is a basic structure built in order to explain how the brain organizes itself to create the rhythmic flow of behavior, including perception and learning. Although the Haken–Kelso–Bunz (HKB) model was originally formulated to account for phase transitions in bimanual movements, it evolved, through experimentation and conceptual elaboration, into a formal construct for the experimental study of rhythmically coordinated movements in general. The model consists of two levels of formalization: a *potential* defining the stability properties of relative phase and a *system* of coupled limit cycle oscillators defining the individual limb movements and their interactions.

Fig. 1. Christiaan Huygens (1629–695).

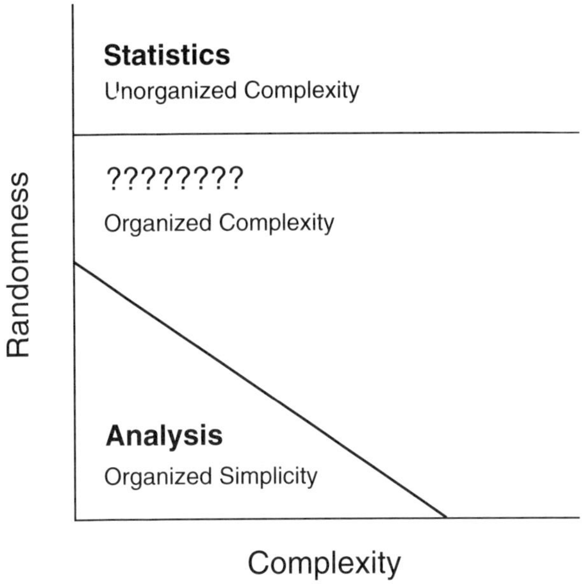

Fig. 2. Complexity, randomness and order.

1.5 Time Dimensions

The time dimension considered in most discussions on dynamical systems is the *absolute* space-time of Newtonian, or classical, physics. In contrast, with this absolute way to look at time-space, Leibniz (1996), who co-discovered differential calculus, had already proposed a *relational* space-time where space is generated by difference and discrimination and time by duration and periodicity.

Rather than thinking of changes as occurring within an independent time flow, we should think of them as *creating time*. We can do this by considering system states as boundaries of time and recurrence patterns as the matter of time (Ward, 2002).

Lorenzana and Ward (1987) argued, in the same line, that systems should be placed in an evolutionary context in which those with increasing complexity emerge from an environment following two principles: *combinatorial expansion* — a linear process whereby the system develops an unrealized potential; and *generative condensation* —

a nonlinear process whereby complexes recombine to create new possible functions. Both processes operate through system-environmental interactions.

Considering all the layers and bindings, waves and cascades of sync phenomena in humans is a great theoretical challenge. This is quite challenging as some of these rhythms are generated at the cellular level as demonstrated by the fact that unicellular organisms express circadian rhythms similar to mammals (Mittag and Volker, 2003). In the individual Central Nervous System (CNS) cells, the expression of so-called clock proteins cycle in double auto-regulatory feedback loops in order to adjust their own transcription in a circadian manner. A circadian rhythm of action potential firing frequency can even be measured in many individual CNS cells when dispersed in cell culture (Welsh, 1995). How then do these individual oscillators communicate with each other in order to construct the resulting *master rhythms*? In the mammalian CNS, the average sum of these individual oscillations can be measured in numerous ways. Studies demonstrate that the circadian rhythmicity of the individual oscillators as well as of the ensemble CNS is maintained even in a brain slice *in vitro* preparation. Given that dispersed cells vary more than cells in organotypic culture, it was deduced that intra-CNS cell coupling is responsible for keeping CNS cells synchronized, even though each autonomous cellular period is determined by its own molecular clock. A model was proposed in which the circadian phases of core oscillators in the CNS are weakly coupled to one another and this sync-core then recruits outlying cells to oscillate under the same period. Eventually, the majority of cells in the CNS oscillate with a period close to the average period expressed by the whole organism (Reppert & Weaver, 2002).

More recently, a study examining brain slice cultures described the CNS as a network of at least three separate groups of oscillators, the phases of which distribute around the average phase of the entire network (Quintero et al., 2003). This phase heterogeneity was hypothesized and could probably arise from intercellular coupling. What sorts of intercellular coupling mechanisms are available to CNS cells? Some coupling, which leads to the expression of some circadian rhythms, may be achieved by also by synapse-independent mechanisms, though direct synapses are responsible for quick and effective interactions. Several alternative coupling pathways have been proposed, such as gap-junctions and diffusible factors.

The discovery that diffusible factors influence circadian rhythms however, does not indicate that diffusible factors are the primary coupling mechanism of circadian rhythms. Diffusible factors and synaptic communication both play a role in coordinating and shaping circadian rhythms. Numerous projections from the CNS to target areas involved in neuroendocrine and autonomic control, are believed to underlie organismic control of circadian rhythmicity (Reppert & Weaver, 2002). Similarly in the CNS itself, diffusible humoral signals, or regularly released peptides, may serve to consolidate the broad circadian rhythms, even in the absence of synaptic neurotransmission. But, the network itself is a circuit of inhibitory synaptic connections that presumably functions as a more immediate cell-to-cell communication mechanism. This may influence other processes as, for example, action potential firing patterns are critical for the coordination between CNS neurons and the projection of a cohesive output pattern to target areas.

1.6 Molecular Oscillators

Independent transcriptional-translational oscillators with relatively short (ultradian) periods can be coupled to generate a circadian oscillator using conventional mechanisms of molecular genetics and reasonable values of parameters describing these mechanisms. The resulting circadian oscillator can be entrained by 24-hour light-dark cycles. The model suggests that evolution of such a circadian oscillator would occur under selective pressure without significantly perturbing the underlying components (Paetkau et al., 2006).

This is a model of a 5-gene circadian oscillator. The components of the first of the primary oscillators are illustrated in the top half of the figure.

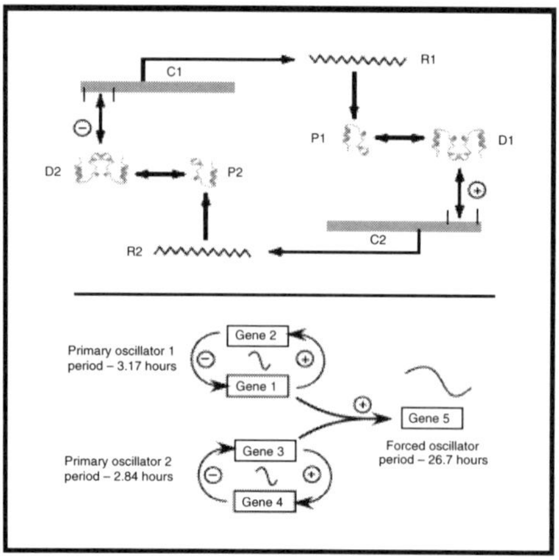

Fig. 3. Model of a gene/protein oscillator (Paetkau et al., 2006).

Fig. 4. Motor synchronies.

The overall model is shown in the lower half of the figure. It comprises two independent, ultradian, primary oscillators (genes 1+2 and 3+4, respectively), in which the homodimeric protein product of gene 1 positively regulates the transcription of gene 2, and a homodimer of protein 2 inhibits transcription of gene 1. Genes 3 and 4 are similarly related. The two primary oscillators differ slightly in their respective periods. The protein products of genes 1 and 3 form heterodimers that regulate the transcription of the fifth gene: the forced oscillator. In the present model, and using the parameters given the periods of the primary oscillators are around three hours, while the period of the fifth gene in the absence of light-dark coupling is just over 26 hours.

1.7 Basic Principles

We will now try to summarize some basic notions on synchronization dynamics in order to prepare our further steps for our exploration on Mind Force. First, we might define *synchronization* as *an adjustment of rhythms of oscillating objects due to their weak interaction.* An *oscillating object*, though this definition might recall pendulums and the origins of sync studies, is *an object displaying a periodic behavior.* The *oscillator* can be *self-sustained*, as an active and autonomous system with an internal source of energy. Alternatively, it may also be referred to as a *forced* oscillator, when the source of energy is external, and it is semi-autonomous. The main dynamical feature of the oscillator is *rhythm*, characterized by *period* (the time spent for each single oscillation) and *frequency* (the number of oscillations per time unit). As we have already seen, if there is more than one oscillator, it is possible for some form of interaction, or *coupling*. The interaction can be mediated by many different media, and is usually weak nor is it easy to recognize or measure. The result of coupling is an adjustment of rhythms, often described in terms of coordination of frequencies; also defined as *entrainment* or *locking* (they are synonyms in sync theory). The *coupling strength* describes how weak or strong the interaction is. Another important feature we might mention is the *tuning* or *detuning* (mismatch): this shows how different two or more oscillators are. If the mismatch is not so...much, there is a higher possibility of entrainment and synchronization. Once sync is nearly in place and we have the possibility that it might be in *phase* or in *antiphase*, like two pendulums that oscillate in the same or opposite directions.

The onset of a certain relationship of two self sustained oscillators will be called *phase-locking*.

There is no synchronization, for example if one system is passive and the other controls it via *resonance*. Another way to exclude synchronization is realizing that the coupling is so strong that there is a *unified system*. In order to categorize a phenomenon as synchronization, we need that all the following criteria are satisfied:

- we analyze the behavior of at least two self sustained oscillators, capable of generating their own rhythms;
- they adjust their rhythms via a weak interaction;
- adjustment of rhythms happens within a certain mismatch.

The conclusion is that synchronization is a complex dynamical process and not a state. Many systems exhibit an alternation of "silence" and rapid activity. These kinds of systems are important in biology as the spiking of neurons, contraction of muscles and macroscopic human behavior can follow this pattern. These kinds of oscillators are called *relaxation oscillators*.

In the 1930s, only periodic self oscillators were known. Nowadays, however, irregular or chaotic self-sustained oscillators are well studied.

We can describe the behavior of the system, as a vector moving by the time evolution of a pair of coordinates. This theater that represents the evolution of the system is called *state* (or *phase*) *space* and the resulting plot can be called *phase portrait*. A periodic oscillation is represented by a closed curve, which can be simple as in the limit cycle or complex as in chaos. The curve, representing all the dynamical behaviors is called an *attractor* and it can be either simple or *strange*, as in the complex quasi-periodicity of chaos (Lorenz, 1993).

The study of synchronization dynamics allows a deeper and clear understanding of how living systems are bounded and connected in the continuous and meandering flow of interactions and resonances we came to call Mind Force.

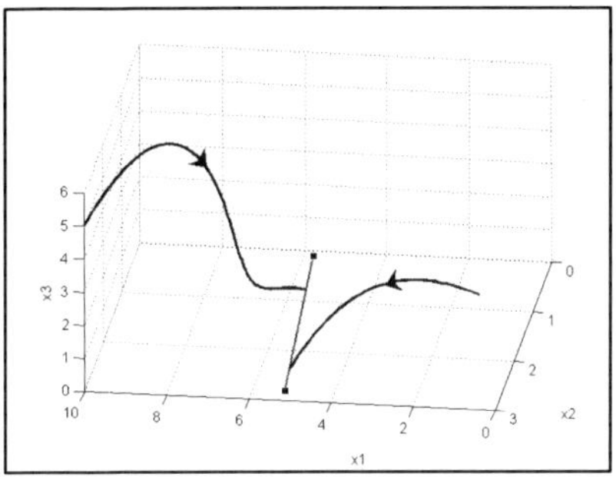

Fig. 5. Line attractor.

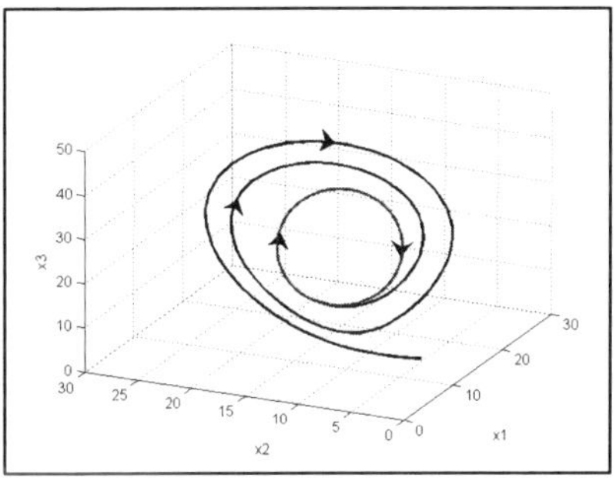

Fig. 6. Circle attractor.

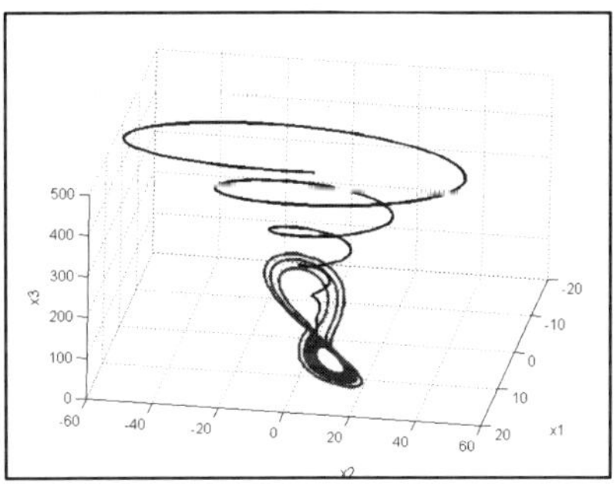

Fig. 7. Strange attractor.

Table 1. Examples of Normal Developmental Coupling of Ultradian Rhythms.

Rhythm	Period, Frequency, Amplitude	Coupling	Comments
heart rate	3 hrs newborn	sleep cycle during development	human
heart rate	3 hrs newborn	circadian 15-30 days of age	human
sleep architecture	newborn, old age	REM 80% REM 20%	human
luteinizing hormone	puberty	sleep + GNRH/LH burst	human: L HR increases 39 fold
nasal cycle	1-5 hrs	autonomic tone and cerebral dominance	decreases with age
blood glucose insulin	6 hrs blood Glucose 24 hrs circadian	mealtime dependent endogenous	healthy adults
Insulin in elderly	irregular release	pulsatile release lost	pulsatile release is lost in elderly

Chapter 2

Consciousness

2.1 Cellular Consciousness

As we have seen oscillators are present almost at every molecular scale of human beings. Consequently, two kinds of questions arise:

- How weak forces produced by oscillators create mass effects via synchronization and recruitment.
- How these mass effects converge in the strands and streams of Mind Force.

Two other subordinate questions are also at stake here:

- How protein and gene timing affect our conscious experience of time.
- If and how this basic form of consciousness impinges on awareness and self-consciousness.

In his paper, "Is Consciousness Only a Property of Individual Cells?" Jonathan Edwards, considers the possibility of recognizing the locus of consciousness in the cell (Edwards, 2005). At the cell level, as we have seen in the previous chapter, there is already a considerable mass of individual molecular oscillators. At the cellular level, there is also the proposal for the role of microtubules and their possible quantum tunneling effects (as we will see more in detail later on). This last contribution to Mind Force might be as molecular as the molecular oscillators we examined, but it would also lead to statistical mechanics and, eventually, quantum dynamics. Nevertheless, Edwards proposes that cellular consciousness "probably does not alter the way we should expect to experience the world, but may help to explain the ways we seem to differ from digital computers and some of the paradoxes seen in mental illness (Orsucci, 1996a). It predicts non-digital features of intracellular computation, for which there is evidence, and should be open to further experimental exploration. The way he depicts cellular consciousness could be considered as an effect of Mind Force fields.

He links consciousness to the notorious *binding problem* in consciousness studies. The binding problem may be defined as the quest of an explanation for the ordered integration of many and varied sensory elements into a single subjective experience. Recent accounts of the importance of this problem to theories of consciousness are given by Chalmers (1995). The problem, actually, is a complex of related problems, two of which we might distinguish, although they are entangled. The first, the *information problem*, is that of the nature of the pathways that bring signals arising at different sites in the brain together as information. The second, the *physical grounding problem* is that of finding a substrate at the fundamental physical level which might support a subjective experience in which many elements are bound into a seamless whole.

The issue of cellular consciousness might be seen as a rather modern one, as resulting from molecular biology discoveries, so it might be a surprise that we find its roots in one of the founding fathers of contemporary psychology: William James.

In Chapter VI of William James's *Principles of Psychology* (James, 1967), he referred to cellular consciousness, as a form of *polyzoism*, the character of being made up of a number of smaller organisms that are acting as a colony. James abandoned single cell consciousness because he was missing some empirical data that we have now found and because he was missing the coherent perspective brought by complexity science and its sub-disciplines (though he had visionary intuitions about some of its basic principles).

We might consider consciousness, as a result of Mind Force within a socio-cultural environment, as a nonlinear scaling property within a biological system. Thus, the key functional requirement of consciousness, as we see it, is that something has simultaneous co-temporal access to many elements of information in defined inter-relationships (i.e. access to a pattern). This requirement is not sufficient to define consciousness but perhaps covers sentience. Consciousness is the kind of sentience in which the accessible pattern includes a useful map of some other 'inner or outer' environment. Three different levels of consciousness can be seen (Orsucci, 2002b):

- consciousness of the wired pattern (sentience);
- consciousness of the outside pattern (a map) and
- consciousness of the self (self consciousness)

These forms of consciousness are all based on binding and synchronization in their various molecular, chemical, electrical and quantum strands. They need to reach, through a nonlinear mass recruitment, different thresholds and scaling.

We agree with Walter J. Freeman (2007) when he writes in a recent paper: "consciousness is not merely 'like' a force; it is a field of force that can be understood in the same ways that we understand all other fields of force within which we, through our bodies, are immersed, and which we, through our bodies, comprehend in accordance with the known laws of physics."

"The aim of consciousness studies should be a set of master equations that relate material and psychosocial state variables, comparable to the equations of Newton, Maxwell and Einstein, but going beyond the physical aspects of the universe and unifying it with the human aspects of the universe. Such a mind theory is needed to provide a scientific foundation for the behavioral sciences, including psychiatry, as envisioned by Freud" (Freeman, 2007).

2.2 Continuous or Discrete

The role of fields, quanta, and mechanics is approached from different fronts by several scientists:

"Time," says Jorge Luis Borges "is the substance I am made of. Time is a river that carries me away, but I am the river...." (Borges, 1964).

Our movements, our actions, are extended in time, as are our perceptions, our thoughts, and contents of consciousness. We live in time, we organize time, and we are time creatures through and through. But is the time we live in, or live by, continuous, as Borges's river? Or is it more comparable to a chain, a series of links or rings connected to or fitted into one another, a succession of discrete moments, like beads on a string? (Sacks, 2004).

David Hume, in the eighteenth century, favored the idea of discrete moments, and for him the mind was "nothing but a bundle or collection of different perceptions, which succeed each other with an inconceivable rapidity, and are in a perpetual flux and movement." (Hume, 1854).

For William James, writing his *Principles of Psychology* the *Humean view*, as he called it, was both powerful and vexing (James, 1967). From the start, it seemed counterintuitive. In his famous chapter on the *stream of thought*, James stressed that, to its possessor, consciousness seems to always be continuous, "without breach, crack, or division", never "chopped up, into bits."

Before 1830 there was no device capable of producing representations or images that had movement. The very idea was paradoxical, a contradiction even. But the *zoetrope* proved that individual images could be fused in the brain to give an illusion of continuous motion, an idea that was soon to give rise to the creation of the motion picture. The zoetrope was an optical toy, in which figures made to revolve on the inside of a cylinder, and viewed through slits in its circumference, appear like a single figure passing through a series of natural motions as if animated or mechanically moved *Zoetropes* and many other similar devices, with a variety of names were extremely popular in James's time, and few middle-class Victorian households were without one. The more recent constructs in consciousness studies mentioning *framing, inner theaters* and *holograms* might be a modern evolution of the zoetrope metaphor.

It is an analogy that Henri Bergson used twenty years later, in his book *Creative Evolution* (1911), where he devoted an entire section to *The Cinematographic Mechanism of Thought, and the Mechanistic Illusion*:

"We take snapshots, as it were, of the passing reality, and...we have only to string these on a becoming, ...situated at the back of the apparatus of knowledge, in order to imitate what there is that is characteristic in this becoming itself.... We hardly do anything else than set going a kind of cinematograph inside us.... The mechanism of our ordinary knowledge is of a cinematographical kind."

Are the brain mechanisms that give coherence to perception and consciousness somehow analogous to motion picture cameras and projectors? Do the eye and brain actually take perceptual stills and somehow fuse them to give a sense of continuity and motion?

2.3 Frames

Oliver Sacks (1999) told us about a rare but dramatic neurological disturbance that a number of his patients have experienced during migraine attacks, when they may lose the sense of visual continuity and motion and see instead a flickering series of "stills". While the effect is somewhat like that of a film (albeit an improperly shot and presented one, in which each exposure has been too long to freeze motion completely and the rate of presentation too slow to achieve fusion), it also resembles some of E.J. Marey's *chronophotographs* of the 1880s, in which one sees a whole array of photographic moments or time frames superimposed on a single plate.

Such visual effects may also occur in certain seizures, as well as in intoxications, especially with hallucinogens such as LSD. Not to mention, there are other visual effects that may occur. Moving objects may leave a smear or wake in the direction in which they move; images may repeat themselves; and after-images may be greatly prolonged. Such standstills showed that consciousness could be brought to a halt, stopped dead, for substantial periods, while automatic functions, maintenance of posture or breathing, for example, continued as before.

Another striking example of a perceptual standstill could be demonstrated with a common visual illusion, that of the *Necker cube*. Normally, when we look at this ambiguous perspective drawing of a cube, it switches perspective every few seconds, first seeming to project, then to recede, and no effort of will suffice to prevent this switching back and forth. The drawing itself does not change, nor does the retinal image. The switching is a cortical process, a conflict in consciousness itself, as it vacillates between alternative perceptual interpretations. This switching is seen in all normal subjects, and can be observed with functional brain imaging.

Fig. 8. A zoetrope advertisement (Circa, 1880).

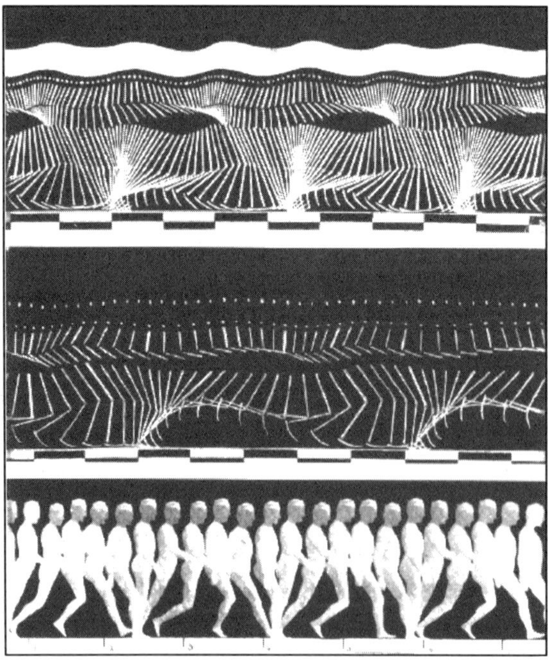

Fig. 9. Étienne-Jules Marey: Cronophotography.

It is similar to what happens when, in a film, the wheels of stage coaches sometimes appear to be going backwards slowly or scarcely moving at all. This wagon-wheel illusion, as it is called, reflects a lack of synchronization between the rate of filming and that of the rotating wheels.

Dale Purves and his colleagues at Duke University (2005) explored wagon-wheel illusions in great detail, and they have confirmed that this type of illusion or misperception is universal. Andrew & Purves (2005) suggest we may find movies convincing precisely because we ourselves break up time and reality much as a movie camera does, into discrete frames, which we then reassemble into an apparently continuous flow.

In addition to physiological studies, there is the relatively new realm of computerized neural modeling, using populations or networks of virtual neurons, and seeing how these organize themselves in response to various stimuli and constraints. All of these approaches, along with concepts not available to earlier generations, now combine to make the quest for the neural correlates of consciousness the most fundamental and exciting adventure in neuroscience today.

A crucial innovation labeled *population thinking*, considers the brain as a neural mass, just as a crowd can produce emerging behaviors different from those of single individuals and push individuals towards behaviors they wouldn't have in other circumstances. Thinking in terms that take the brain's huge population of neurons (a hundred billion or so) into account, one can realize the power of experience that differentially alters the strengths of connections between them and promotes the formation of functional groups or constellations of neurons throughout the brain. These groups and their interactions serve to categorize experience.

Instead of seeing the brain as rigid, fixed in mode, programmed like a computer, there is now a much more biological and powerful notion of *experiential selection* of experience: literally shaping the connectivity and function of the brain (within genetic, anatomical, and physiological limits, of course). Such a selection of *neuronal groups* (groups consisting of perhaps a thousand or so individual neurons), and its effect on shaping the brain over the lifetime of an individual, is seen as analogous to the role of natural selection in the evolution of species; hence Gerald M.

Edelman, who was a pioneer in such thinking in the 1970s, speaks of *neural Darwinism* (Darwin, 1909; Edelman, 1989). J.P. Changeux (2004), the French neuroscientist, is more concerned with the connections of individual neurons, and speaks of *the Darwinism of synapses*. While Walter J. Freeman focuses on global dynamics and speaks of *neural masses* and *fields* (1975).

William James himself always insisted that consciousness was not a *thing* but a dynamical process. The activity of a coalition of neurons, or coalition of coalitions, if it is to reach consciousness, must not only cross a threshold of intensity, but must be held there for a certain time; roughly a hundred milliseconds. This is the duration of a *perceptual moment*, resounding of the *specious present* mentioned by James and the επoκη (the duration of frames has to be defined 100-300 ms or even 7 seconds)

To explain the apparent continuity of visual consciousness, Crick and Koch (2003) suggested that the activity of the coalition shows *hysteresis*, that is, a persistence outlasting the stimulus. This notion is very similar, in a way, to the *persistence of vision* theories advanced in the nineteenth century. In his *Physiological Optics* (1860), Hermann Helmholtz wrote, "All that is necessary is that the repetition of the impression shall be fast enough for the after-effect of one impression not to have died down perceptibly before the next one comes". Helmholtz and his contemporaries supposed that this aftereffect occurred in the retina, but for Crick and Koch it occurs in the coalitions of neurons in the cortex.

Vision, in ordinary circumstances, is seamless and gives no indication of the underlying processes to which it depends on. It has to be decomposed, experimentally or in neurological disorders, to show the elements that compose it. Thus it is decomposed vision, the flickering, perseverative, time-blurred images experienced in certain intoxications or severe migraines, which above all lends credence to the notion that consciousness is composed of discrete moments.

If a dynamic, flowing consciousness allows, at the lowest level, a continuous, active scanning or looking, it allows, at a higher level, the interaction of perception and memory, of present and past. And such a *primary consciousness*, as Edelman puts it, is highly efficacious, highly adaptive, in the struggle for life.

From such a relatively simple primary consciousness, we leap to human consciousness, with the advent of language and self-consciousness and an explicit sense of the past and the future. And it is this which gives a thematic and personal continuity to the consciousness of every individual.

2.4 Mindfulness

We have been exploring how the result of massive cascades of synchronizations that we proposed to call Mind Force is producing the internal biophysical cohesion which is preliminary to any conscious thinking process. We decided to call the result of MF at this level as *Biophysical Self.* There are ways in which this Self finds conscious expressions in thinking, social relations and culture. We already explored some perspectives on this. Now we might consider another perspective on MF. An important conscious expression of reflexivity and mirroring dynamics is what we call *empathy* (from the Greek "feeling with") or *attunement* to express a way a person can be in tune with the internal world of another (a parent with a child and vice versa, lovers etc.). This area of studies is connected also with what has been called *emotional intelligence* (Goleman, 1995).

As we have seen in previous parts of the book, there is a neural circuitry that enables people to feel by each other. "This state is crucial if people in relationships are to feel vibrant and alive, understood and at peace. Research has shown that such attuned relationships promote resilience and longevity" (Siegel, 2007: xiv). Mindfulness is the same process happening within the subject, an inner attunement which enables integration in thinking and neural integration as well, flexibility and self-understanding: "being mindful is a way of becoming your own best friend" (Siegel, 2007: xiv). This is a way to use the stream of Mind Force in order to promote physical and psychological well being with mindful awareness.

John Kabat-Zinn (2003: 144) has proposed an effective definition of mindfulness as "the awareness that emerges through paying attention on purpose, in the present moment, and nonjudgmental to the unfolding of experience moment by moment." The contribution of religious traditions to the development of this capacity has been important in the East (Buddhism, Yoga, and Taoism) and in the West, where Christian, Muslim and Jewish traditions of mindful practice have always been active, though less evident (Goleman, 1977; Schopen & Freeman, 1992).

The main psychological functions in place might be named as in a review of studies (Baer et al., 2006):

- non reactivity;
- observing;
- non acting or acting with awareness;
- describing the experience;
- no judgment.

As Siegel notes, "being mindful opens the doors not only to be aware of the moment in a fuller way; but by bringing the individual closer to his or her inner world, it offers the opportunity to enhance compassion and empathy (Siegel, 2007). Clinical applications of mindfulness have focused on pain relief, substance abuse, personality disorders, obsessive compulsive disorder, post-traumatic stress disorder, ADHD and other conditions (Siegel, 2007). Mindfulness-based interventions do not focus on achieving cognitive or behavioral change; however, such change is often a consequence of being mindful.

Sometimes mindfulness is defined with slightly different terms as mentalisation or reflective function but the basic psychological and neurophysiologic processes are the same. These last definitions stress the connections between mindfulness and attachment research which we might explore further in the next chapter.

Mindfulness approaches, unlike traditional psychotherapy, do not directly address the content of thought but rather alter the individual's relationship to their thoughts. As described earlier, mindfulness involves the nonjudgmental observation of constantly changing internal and external stimuli as they arise, and this skill is developed through a variety of meditation exercises. Mindfulness-based tasks encourage the individual to attend to either internal experiences, such as bodily sensations, thoughts and emotions, or external stimuli, such as sounds.

The individual is required to observe these experiences but not evaluate them or react to them. Baer asserts that mindfulness involves 'flexible awareness'. Individuals are asked to focus on the target of observation in the present moment, for example, body sensations or breathing, acknowledging passing thoughts and sensations, but not focusing on their content.

2.5 Brain Changes

The nervous system develops from one of the main three parts of the embryo we call *ectoderm*, the same outer part from which develops the skin. At some stage of development, certain clusters of these outer cells fold inward to form the neural tube and the spinal chord. The brain originates at the interface of the inner and outer worlds, as the topmost part of the nervous system distributed throughout the body. When we use the word *embodiment* we should remember that it is not just a psychological concept but also the result of a developmental process and of the architecture of our bodies.

The inner neural structure and network of synapses is produced by the genetic material and the interactions with environments, including relationships. Neuroplasticity is the way we call the continuous remolding of the nervous system: it has been estimated that everyday about the 70% of our synapses can modify their functioning and/or their connections. There are many different ways we can influence these changes, and a powerful one is meditation and mindfulness.

There is a quantity of studies accumulating evidence on brain changes induced by meditative and mindfulness states. For example, Lazar et al. (2005) used magnetic resonance imaging to assess cortical thickness in 20 participants with extensive Insight meditation experience, which involves focused attention to internal experiences. Brain regions associated with attention, interoception and sensory processing were thicker in meditation participants than matched controls, including the prefrontal cortex and right anterior insula. Between-group differences in prefrontal cortical thickness were most pronounced in older participants, suggesting that meditation might offset age-related cortical thinning.

Finally, the thickness of two regions correlated with meditation experience. These data provide the first structural evidence for experience-dependent cortical plasticity associated with meditation practice.

These long lasting changes in brain structures are preceded by specific functional changes during the meditative experience. In a study by Aftanas & Golosheikin (2003) EEG spectral power and coherence estimates in the individually defined delta, theta, alpha-1, alpha-2, and alpha-3 bands were used to identify and characterize brain regions involved in meditative states, in which focused internalized attention gives rise to emotionally positive 'blissful' experience. Blissful state was accompanied by increased anterior frontal and midline theta synchronization as well as enhanced theta long-distant connectivity between prefrontal and posterior association cortex with distinct 'center of gravity' in the left prefrontal region (AF3 site). Subjective scores of emotional experience significantly correlated with theta, whereas scores of internalized attention with both theta and alpha lower synchronization. Our results propose selective associations of theta and alpha oscillating networks activity with states of internalized attention and positive emotional experience.

Some brain areas are more involved than others during meditative states. In a study on functional brain mapping of the relaxation response and meditation, Sara Lazar et al. (2000) used functional magnetic resonance imaging (fMRI) to identify and characterize the brain regions that are active during a simple form of meditation. It was hypothesized that neural structures that have a role in attention and arousal would be activated during meditation and that the fully developed relaxation response would differ from the early stage of meditation. Five 22-45 yr olds participated. Significant signal increases were observed in the group-averaged data in the dorsolateral prefrontal and parietal cortices, hippocampus/parahippocampus, temporal lobe, pregenual anterior cingulate cortex, striatum, and pre- and post-central gyri during meditation. Global fMRI signal decreases were also noted, although these were probably secondary to cardiorespiratory changes that often accompany meditation. Results indicate that the practice of meditation activates neural structures involved in attention and control of the autonomic nervous system.

2.6 Traditions

It seems that religious traditions from shamanism to modern church chanting, have an empirical knowledge of the powers and efficacy of mindfulness. For example, researchers at the University of Wisconsin–Madison (Austin, 2006) have found that during meditation, Zen Buddhist monks show an extraordinary synchronization of brain waves known as gamma synchrony.

Most conscious activity produces beta waves at 13 to 30 hertz, or cycles per second. More intense gamma waves (30 to 60 or even 90 Hz) generally mark complex operations such as memory storage and sharp concentration. The Wisconsin study took electroencephalograms (EEGs) of 10 longtime Buddhist practitioners and of a control group of eight college students who had been lightly trained in meditation. While meditating, the monks produced gamma waves that were extremely high in amplitude and had long range gamma synchrony — the waves from disparate brain regions were in near lockstep, like numerous jump ropes turning precisely together. The synchrony was sustained for remarkably long periods, too. The students' gamma waves were nowhere near as strong or tuned. Such results connote more than spiritual harmony; they reflect the coordination of otherwise scattered groups of neurons. Gamma synchrony increases as a person concentrates or prepares to move. And lack of synchrony indicates discordant mental activity such as schizophrenia.

A growing body of theory proposes that gamma synchrony helps to bind the brain's many sensory and cognitive operations into consciousness. That hypothesis certainly agrees with the monks' gamma readings, seemingly confirming that Zen meditation produces not relaxation but an intense though serene attention. Trained musicians also show superior gamma synchrony while listening to music — another form of calm but intense focus.

Sona Dimidjian & Marsha Linehan (2003) posed several very useful questions for the development and future applications of mindfulness. What are the consequences of separating mindfulness from its spiritual and cultural origins? Is mindfulness training an efficacious treatment intervention? What are the active or essential ingredients of mindfulness training? Can mindfulness enhance clinical practice apart from its role as a clinical intervention? How does mindfulness work? How should therapists be trained in order to deliver mindfulness interventions competently? Is mindfulness training amenable to widespread dissemination?

Steven Hayes (2003) raised a very interesting question about the clinical and social impact of a larger diffusion of mindfulness practice. "Understanding the processes and principles that underlie mindfulness is a needed step, because this method enters into the armamentarium of empirical clinical psychology. Mindfulness is closely related to several procedures, including acceptance, cognitive defusion, and exposure. Although each of these procedures seems to target different behavioral processes, they are all interrelated, because ultimately all of them target the domination of the literal and evaluative functions of human language and cognition. Because these methods are constructional and inclusive, not eliminative or exclusive, their rise may ultimately have a more profound impact on the field than is currently supposed."

2.7 Semantic Networks

At the cognitive level we recognize the imprint of Mind Force hyperstructure in other ways. Charles Sanders Peirce, in his correspondence with her, approved Lady Welby's dicta that "language is only the extreme form of expression" and that "life itself may be considered as the Expression". We want to recall something from his letter of October 12, 1904:

"Any concept is a sign, of course. Ockam, Hobbes and Leibniz have sufficiently said that. But we may take a sign so broad a sense that the interpretant of it is not a thought, but an action or experience, or we may even so enlarge the meaning of sign that its interpretant is a mere quality or feeling. A Third is something which brings a First into relation to a Second. A sign is a sort of a Third. How shall we characterize it? Shall we say that a Sign brings a Second, its Object, into cognitive relation to a Third? ...If we insist on consciousness, we must say what we mean by consciousness of an object" (Peirce, 1931).

He was referring to his catalogue of semiotic entities: "I divide signs into Icons, Indices and Symbols..." In this catalogue there were two basic connected triads to define the semiotic universe. One triad is defining typologies of signs; the other one is defining relations between semiotic event, reference and other signs.

This definition of the basic semiotic unit generates chains of relationships and geometries in the Semiotic Universe This can give an idea of the possible landscape of a small area of the Semiotic Universe, based on Peirce's view (Orsucci, 1981, 2002; Eco, 1989, 1990). The first representations of the Semiotic Universe are the *arbor porphyriana* or Porphyrian tree, created by Porphyry of Tyre (A.D. 233–c. 309), a Phoenician Neoplatonic philosopher. Porphyry's *Isagoge*, or Introduction to Aristotle's *Categories*, in Latin translation, was the standard textbook on logic for at least a millennium after his death (1940). A *Porphyrian Tree* is a hierarchical (tree structured) ontology, construction in logic consisting of three rows or columns of words; the middlemost whereof contains the series of genus and species, and bears some analogy to the trunk. The extremes, containing the differences, are analogous to the branches of a tree.

Most modern representations of the Semiotic Universe are semantic networks. A Semantic network is a network, which represents semantic relations between the concepts. This is often used as a form of knowledge representation. It is a directed or undirected graph consisting of vertices, which represent concepts, and edges. Semantic Nets were first invented for computers by Richard H. Richens of the Cambridge Language Research Unit in 1956 as an *Interlingua* for machine translation of natural languages.

They were developed by Robert F. Simmons at System Development Corporation in the early 1960s and later featured prominently in the work of Ross Quillian (1967).

In the 1960s to 1980s the idea of a semantic link was developed within hypertext systems as the most basic unit, or edge, in a semantic network. These ideas were extremely influential, and there have been many attempts to add typed link semantics to HTML and XML.

The trajectories of meaning percolation within the Semiotic Universe can be represented, for example, in a quite effective way by TextArc, a software designed to help people discover patterns and concepts in any text by leveraging a powerful, underused resource: human visual processing. It compliments approaches such as Statistical Natural Language Processing and Computational Linguistics by providing an overview, letting intuition help extract meaning from an unread text. Here, an analysis of Lewis Carroll's Alice in Wonderland demonstrates TextArc's structure and some capabilities (Paley, 2002).

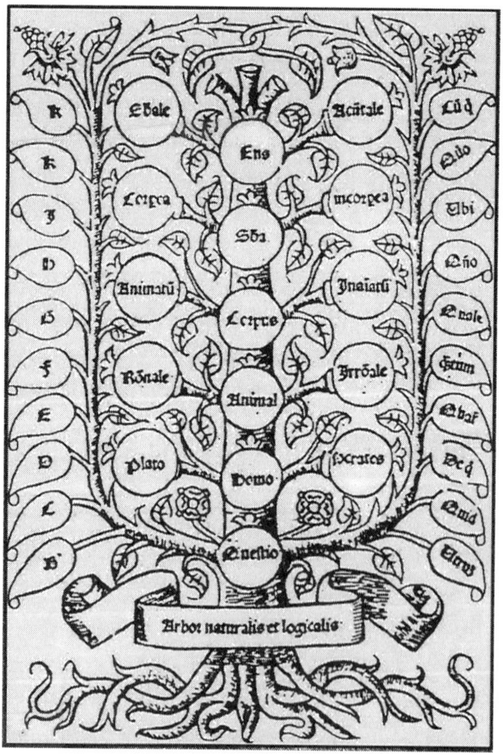

Fig. 10. Porphyrian tree.

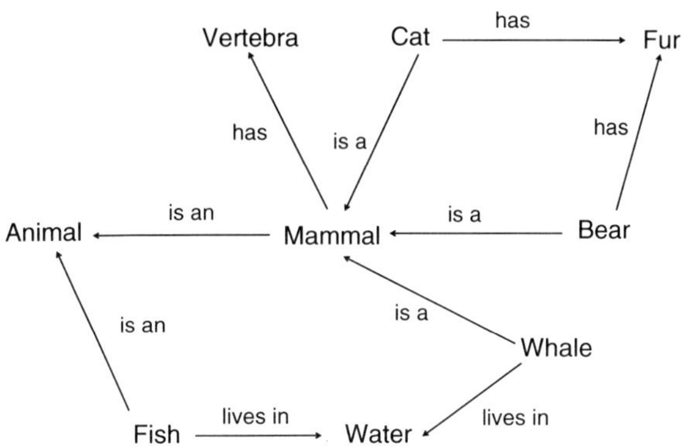

Fig. 11. Example of semantic network.

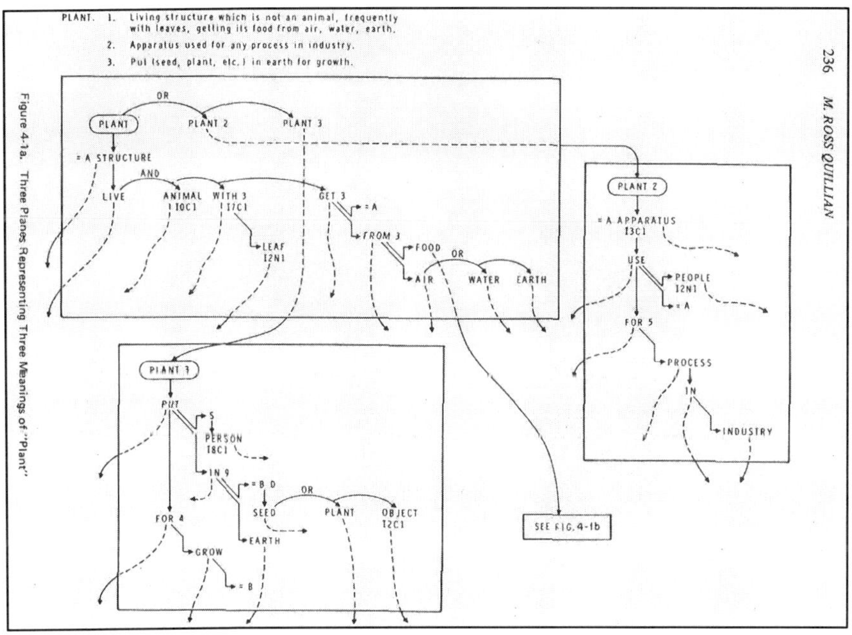

Fig. 12. Meaning percolations in semantic networks (Quillian, 1967).

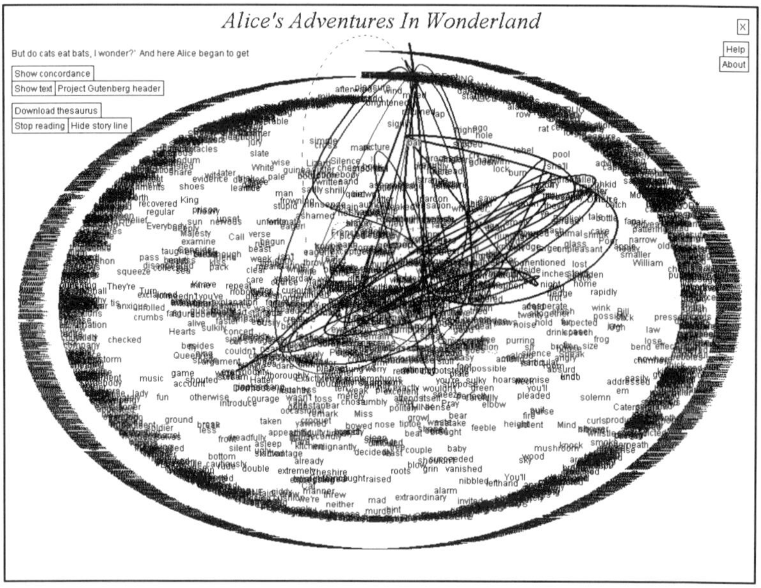

Fig. 13. Semiotic trajectories, Alice in TextArc (Paley, 2002).

Chapter 3

Substratum

3.1 Beyond Descartes

One of the outstanding problems in cognitive sciences is to understand how ongoing conscious experience is related to the workings of the brain and nervous system.

Neurodynamics offers a powerful approach to this problem because it provides a coherent framework for investigating change, variability, complex spatiotemporal patterns of activity, and multiscale processes, among others. We advocate for a neurodynamical approach to consciousness that integrates mathematical tools of analysis and modeling, sophisticated physiological data recordings, and detailed phenomenological descriptions. We begin by stating that consciousness is an intrinsically dynamic phenomenon and must therefore be studied within a framework that is capable of rendering its dynamics intelligible. Dynamics in a narrow sense refers to the change any system goes through in time. In a broad and modern sense, it refers to nonlinear dynamical systems, including mathematical models and empirical studies (Van Gelder, 1998). Finally, it refers to the phenomenology of our conscious experiences and the way they unfold during time (Varela et al., 1991; Varela, 1999).

These different domains are and should be considered as intermingled in order to respect the intrinsic complexity of the living, despite all difficulties to explain and understand it (Reda, 1986; West, 2006). Traditional ways to think of causal relations can be challenged by the high sensitivity to initial conditions. For example, this property of complex systems makes the application of brute force inappropriate in case a change is required. A succession of carefully chosen and delicate inputs is always more appropriate. One could use the metaphor of trying to catch a fish out of water (Pecora & Carroll, 1990; Boccaletti, Pecora, & Pelaez, 2001).

Living systems contain an intrinsic variability that cannot be attributed just to noise or other environmental conditions, but appear constitutive of their functioning. Moreover, sometimes the internal architecture of dynamical systems reveals a balance of segregated and cooperating subsystems, based on the transient and distributed coupling of their "parts".

Some of these systems also display what has been referred to as self-organization, as the emergence of collective coherent behavior starting from random initial conditions. This feature has been proven to be fundamental in biological phenomena (Haken, 1983; Kelso, 1995).

Mind is a complex dynamical structure in which a cooperation of neurons, the whole body, behavior, language and sociocultural conditions produce a coherent identity. Mind is never silent but always functioning in a multiplicity of different states, somewhat recurring and always changing (Orsucci, 2002b). Something William James (maybe implicitly recalling Heraclitus) called *stream of consciousness*. There is always something old and at the same time something new, even unpredictable. The convergence of complex dynamical patterns in experience and brain activity suggests that there is the intrinsic hyper-coordination that we came to call *Mind Force*.

Neurodynamics emerged from a proposal which can be traced back to the 50s (Ashby, 1952) when the nervous system was recognized as a nonlinear dynamical system. More recently, these statements have been extended by the recognition of the so-called chaos in the brain. This recognition, started by Walter Freeman and other important researchers, who started a debate about whether CNS is actually chaotic and what use could make CNS of chaos (Skarda & Freeman, 1987; Tsuda, 2001). We might extend this last question further by asking what evolutionary advantage could there be in the usage of chaotic dynamics by CNS. For example, in olfaction we have seen that if an animal is confronted with slightly different olfactory stimuli, the dynamic trajectory of cortical activity will converge onto the same attractor. Thus in this model, perception is based on several coexisting attractors in a multistable system. Thus, it is not yet clear if the whole activity of the nervous system is chaotic as the behavior of the nervous system which cannot be observed directly, as it has been done *in vivo* on some specific areas. The devices we use to observe CNS are crude projections of its actual state just like the EEG, SPECT, MRN (and other techniques designed by even more arcane acronyms).

Nevertheless, the debate over the chaotic nature of brain dynamics brought some useful ideas such as considering the variable and nonlocal neuronal activity as an integral part of brain dynamics. For this reason, the neurodynamical approach is often presented as a sharp alternative to the computer metaphor of the brain (Freeman, 1999a; Kelso, 1995; Van Gelder, 1998).

As Crutchfield and Kaneko (1987) note, dynamical system theory has developed largely through the study of low-dimensional systems, with no spatial extension. To be useful for neuroscience, however, dynamical system theory needs to consider the special properties conferred on the nervous system by its spatial extension. As a result, there is now a shift of interest on the properties of self-organization and the formation of transient structures of the brain. As noted by Freeman in the preface of the second printing of his seminal book *Mass Action in the Nervous System* (Freeman, 1975): "The word chaos has lost its value as a prescriptive label and should be dropped in the dustbin of history, but the phenomenon of organized disorder constantly changing with fluctuations across the edge of stability is not to be discarded" (Freeman, 2004).

In his pioneering work, Freeman observed spatiotemporal activity patterns in the olfactory bulb and interpreted them within the framework of Non Linear Dynamics. Afterwards, in an influential theoretical paper with Christine Skarda, he proposed that sensory information was *encoded* in those patterns (Skarda & Freeman, 1987). The classical example of spatiotemporal structures in the brain is the Hebbian reverberant cell assembly which Hebb (1949) hypothesized as a basis for short term memory (Amit, 1989). Reverberant cell assemblies are transitory sets of neurons that oscillate together at the same frequency. They are the best studied spatiotemporal structures in the brain. The cortex has been modeled as a mesh of coupled oscillators (i.e. a lattice of reverberant cell assemblies). The formation of these structures often takes the form of phase synchronization patterns between oscillators expressed on the macroscale as EEG recordings and on the mesoscale as local field potentials. A reason why resonant cell assemblies are so appealing to neuroscience is that they provide a flexible and reversible way to bind together distributed neurons, maybe involved in very different functional processes. This type of binding has three fundamental features:

- *integration* of distributed neural activities;
- *segregation* of particular sets above the rest of the brain and
- *metastability* of patterns evolving through a succession of changes.

Francisco Varela and others proposed that every cognitive act corresponds to the formation of such a transient spatiotemporal pattern. This approach can be a heuristically useful scheme to be generalized to transient patterns integrating neurodynamics with behavioral and cognitive coordinations.

Walter Freeman, after his studies on neurodynamics, has been trying to extend the three fundamental features of integration, segregation and metastability to the psychosocial domains of attachment dynamics, social patterns etc. We will try to go deeper in these applications of self organization and transients in the next chapter. For Freeman *meaning is embodied* in these amplitude-modulation patterns of neural activity, whilst structure is dependent on synaptic changes. The whole story of an animal is embedded in meaningful spatiotemporal patterns. In Freeman's view, perception is essentially an active process closer to hypothesis testing more than to passive information recovery. It is an active stance based on the intentional and attentive expectancy. Stimuli confirm or disconfirm these hypotheses through state transitions generating new amplitude modulation patterns. The multisensory convergence generates landscapes, forms, *Gestalten*, as discrete structures in the action/perception process.

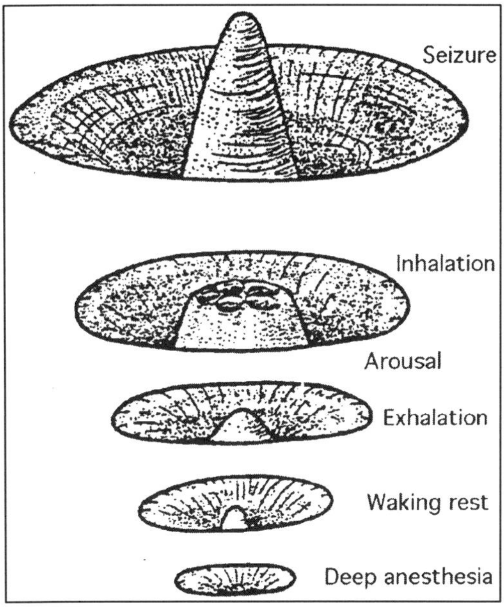

Fig. 14. Mesoscopic neurodynamics (Freeman, 2000).

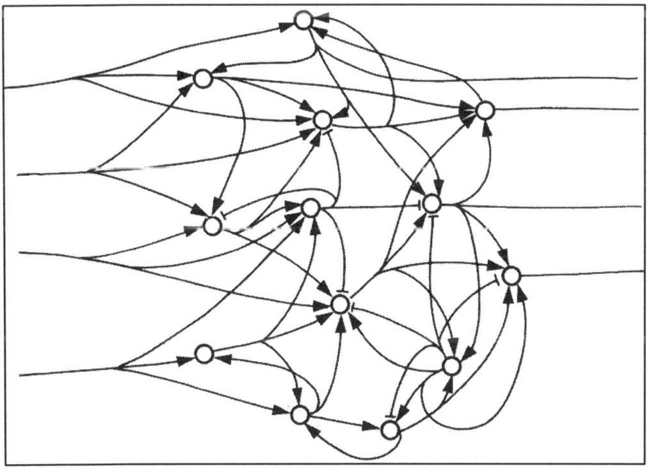

Fig. 15. Neuronal assembly with excitatory and inhibitory connections.

On a quite parallel line of thinking, Francisco Varela & Evan Thompson (2001) analyzed "the flow of adapted and unified cognitive moments." They showed how a discrete cognitive act is linked to a transient and distributed neural assembly. Moreover, the relevant variable is the dynamic nature of links established between different neurons and different sets of them. Phase synchronies are mediating such dynamical links, especially in the beta (15-30 Hz) and gamma (30-80 Hz) frequencies. Varela and Thompson (2001) qualified their perspective as *radical embodiment* emphasizing that large scale and transient space/time patterns are crucial. Yet what's even more important is the way they dynamically integrated these brain dynamics with body/environment dynamics and in brain-body-world cycles of operation (Thompson et al., 2001).

The inner organismic integration of sensors and effectors is *core consciousness* (Damasio, 1998b, 1998a) or *primary process consciousness* (Panksepp, 1998). The sensory-motor coupling between organism and environment is coevolving process, all along with the intersubjective action of behavioral and linguistic communication. Here, mirror systems bring reflexive and reverberating cascades posed as junctions between coevolving processes (Rizzolatti & Arbib, 1998).

Edelman & Tononi (2000) stress that consciousness is not a thing but a process and therefore should be explained just in terms of processes and interactions rather than as maps of local activity. Integrations and differentiation, in their view are important at the same in their "*dynamic core* hypothesis." In order to sustain a conscious experience, in their view, a functional cluster of neurons must be highly differentiated and segregated by functional borders, based on the timing of their activity. This is the reason why the dynamic core of neural functions is constantly moving and changing: it "is not a thing or a place" on a cortex map.

The *dynamic core* is probably something more than what Edelman and Tononi described. It can be considered as a *dynamical matrix* from which mind and matter emerge as seemingly separate entities. From a synchronization and network theories point of view, it can be seen as a *hyperstructure,* a network of networks of oscillators. This is what we are now calling *Mind Force*. The examination of some attempts to define consciousness in terms of Quantum Theory may shed more light on this.

3.2 Waves And Quanta

Ilya Prigogine was one of the great visionaries of the 20th century, and a Nobel Prize winner in 1997. His work on non-equilibrium thermodynamics, dissipative structures, complexity science and the arrow of time are now landmarks in the history of science. Prigogine encouraged many kinds of interdisciplinary work and, late in his life, when asked about the future challenges of science, he answered: "If I was a young researcher now, I would study the mind-body problem. This is the great challenge of the 21st century." Aside from traditional human sciences as philosophy and psychology, we have seen that dynamical systems theory has something remarkable to say on this matter. We will see later in this book how network theory has dealt with these problems.

Quantum physics has been used as another, not necessarily alternative, perspective and some using this paradigm have claimed this could be the real key to solve the problem. An examination of several quantum approaches will be useful to understand something more on the nature of Mind Force.

As Harald Atmanspacher (Atmanspacher, 2004) specified in his clear and thoughtful critical review of the issue: "Although there can be no reasonable doubt that quantum events happen in the brain as elsewhere in the material world, it is the subject of controversy whether these events are in any way *efficacious* and *relevant* for those aspects of brain activity correlated with mental activity." It is important to remember that the original motivation for the pioneers of quantum theory in the 20th century to approach mind studies was basically philosophical. It was related to the introduction of randomness standing out against the previous deterministic worldview. Complementarity and entanglement are other concepts relevant for consciousness studies, considered by the pioneers of quantum physics such as Planck, Bohr, Schrödinger and Pauli.

In its most general form, the mind-matter distinction comprises not only the distinction between mind and body but also, even more specifically, that of mind and brain. Such dichotomies have been posed in both epistemological and ontological frameworks, and elaborated in quite a number of variants. They range from a fundamental distinction of mind and matter at a primordial level of description to the emergence of mind (consciousness) from the brain as an extremely sophisticated and highly developed material system.

Correlation versus causation is one of the key issues in all discussions on the mind/matter problem. It is not difficult, as we know; pointing out correlations between mental and material states (e.g. neuromediators, hormones, electric potentials etc.). It is by far more difficult to prove a causal feature of these correlations. The passage from descriptions to explanations and theories is often difficult and risky in this area. In physics there are four established kinds of fundamental interactions: electromagnetic, weak, strong and gravitational. However, in this area it is always difficult to explain the nature or even the existence of interactions: "the existing body of knowledge basically consists of empirical correlations between material and mental states" (Atmanspacher, 2004). In some areas such as pharmacology or brain imaging there are correlations in which a *post hoc propter hoc* has a probabilistic value, but there is not yet a coherent theory about this.

As Atmanspacher proposes, in most approaches used to explain material brain states and mental states, relations are explained as reciprocal and direct, providing a minimal framework for reduction, supervenience and emergence. As an alternative view, "it is possible to conceive mind-matter relations indirectly via a third category" (Atmanspacher, 2004). In the early days of psychophysics, during the 19th century, Fechner and Wundt advocated similar views. More recently Whitehead (1978), the modern pioneer of process philosophy, referred to the mental and the material as "actual occasions" and appearances. Other similar variants of the same approach have been proposed by Pauli & Jung (1955) or Bohm & Hiley (2008), mentioning an *implicate order* which unfolds in different domains. Between contemporary authors Chalmers (1996) proposes that the neutral level of description could be defined in terms of *information*. In modern cognitive science, however, the mental and the material views are sometimes defined in terms of first person (including *qualia*) and third person perspectives.

We have already dealt with the question "what is the neural correlate of a mental representation?" (also called *grounding* problem). From the neurodynamical point of view neuronal assemblies (ensembles of thousands coupled neurons) are the answer. Each assembly is usually *distributed*, not spatially contiguous but rather dynamically *segregated*: inner connections, within each assembly, are stronger than connections from the outside. In order to achieve an operational stability the balance between positive and negative feedback is also important. Most of grounding models, as we have seen, do not involve quantum theory, but some exceptions are remarkable and can help to further clarify the nature of Mind Force.

Before going deeper in the examination of each of the proposed quantum approaches, a *caveat* should be expressed. There are a number of accounts discussing quantum theory in relation to consciousness in a *metaphorical* way. In metaphorical terms, basic concepts of quantum theory such as entanglement, superposition, collapse, complementarity and others are used for their evocative power.

Non-metaphorical quantum approaches are quite different though explicit contaminations between metaphors and formalized scientific ideas frequently happen and can be a source of creative thinking and heuristics (Verhulst, 1994; Orsucci, 2004). For example, conscious acts can be just interpreted as analogies of physical measurements, or correlations in psychological terms can be considered as analogues to physical entanglement. "Unless this detailed work leads beyond pure metaphors and analogies, they do not yet represent scientific progress" (Atmanspacher, 2004).

A second category includes state of the art quantum theory used to explain current neuroscience issues. A line of research started by von Neumann (2004) considers quantum state reductions as conscious acts. As we have seen, the act or process of measurement is a crucial aspect of quantum theory and von Neumann considered it as a collapse or reduction of the wave function. He differentiated three functionally separate entities: the object system, the measuring device and the observer. The (subject/object) boundary, in his opinion, could be positioned between any of these three entities and the brain could be either a subject or an object in the measurement process. Stapp (1993) used the freedom to place the interface by placing it within the observer's brain, in this way splitting the subject. He argued that quantum uncertainties at the synaptic level can produce effects large enough to generate superpositions of macroscopic patterns of brain activity at the level of neuronal assemblies. The neural correlate of conscious events is then assumed to be the collapse of such superposition into an actualized neuronal assembly. Mental states, in this way, may produce mind/matter causal interactions, corresponding to quantum collapses of superposition states in brain activities.

In the 60s, Ricciardi & Umezawa (1967), suggested to utilize the formalism of quantum field theory to describe brain states, with particular emphasis on memory. They proposed to conceive memory states as many particle systems, representations equivalent to vacuum states in quantum fields. Pribram (1971) in the same years proposed similar views, a reducibility of mental activity to brain activity, by considering the brain as a multiple particle system, where the "particles" are neurons.

Beck & Eccles (1992) applied quantum mechanics to describe the process of *exocitosis*, the release of neurotransmitters in inter-synaptic space, and a chemical component of neurotransmission regulated by ion (mainly calcium).

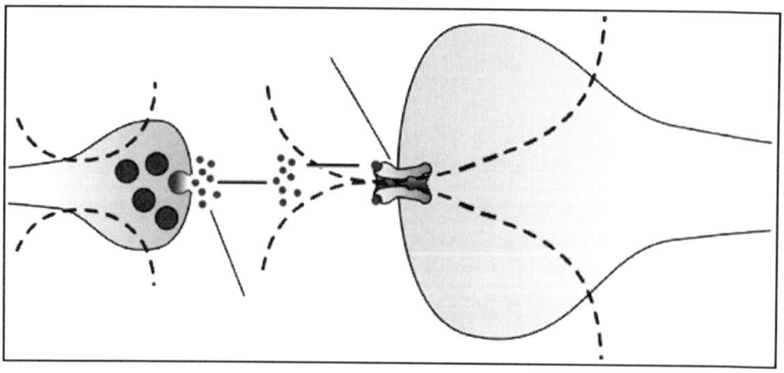

Fig. 16. Release of neurotransmitters at the synaptic cleft (exocitosis).

Finally, another cellular "department" in which quantum processes have been proposed is the cytoskeleton. It consists of protein structures made of filaments and microtubules. Microtubules are polymers arranged in a tubular array, with an outside diameter of 25 nm. Roger Penrose in his first book on consciousness, The Emperor's New Mind (1989), based

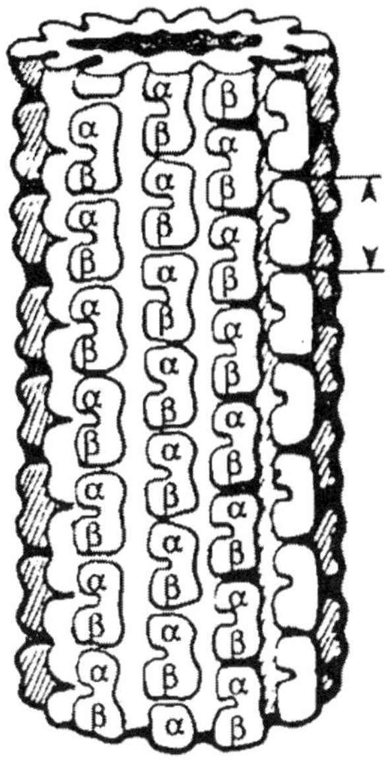

Fig. 17. Microtubule and its protein structure.

on Gödel's incompleteness theorems, argued that the brain could perform functions that no computer or system of algorithms could. Consciousness might be fundamentally non-algorithmic, and incapable of being modeled as a classical Turing machine type of computer (Turing, 1996). Penrose saw the principles of quantum theory as providing an alternative process through which consciousness could arise. He further argued that this non-algorithmic process in the brain required a new form of *quantum wave reduction*, later given the name *objective reduction*, which could link the brain to the fundamental spacetime geometry. At this stage, he had no precise ideas as to how such a quantum process might be instantiated in the brain. Moreover, Penrose's ideas were widely criticized by neuroscientists, logicians and philosophers, notably Grush & Churchland (1995).

Stuart Hameroff was inspired to contact Penrose regarding his own theories about the mechanism of anesthesia, and how it specifically targets consciousness via action on neural microtubules. Penrose was interested in the mathematical features of the microtubule lattice, and the two collaborated in formulating a so-called *orchestrated objective reduction* (Orch-OR) model of consciousness. Following this collaboration, Penrose published his second consciousness book, Shadows of the Mind (1994).

This more developed version of their ideas was also widely attacked, notably by the physicist, Max Tegmark, who calculated that quantum states in microtubules would survive for only 10^{13} seconds, too brief to be of any significance for neural processes (Tegmark, 2000). Hameroff and the physicists, Scott Hagan and Jack Tuszynski (Hagan, Hameroff & Tuszynski, 2002) replied to Tegmark arguing that microtubules could be shielded against the environment of the brain. To date, there is no experimental confirmation of these proposed methods of shielding, but Hameroff has proposed tests that could falsify the theory.

3.3 Synchronization

As seen in the previous chapter, we can study these dynamical behaviors in terms of coupling and synchronization. Synchronization of coupled oscillating systems means the appearance of certain relations between their phases and frequencies. It can be used to modify a system by controlled interaction, and also, in the reverse sense to detect some features of the coupling structure. Synchronization can happen between periodic, noisy or chaotic systems. Recently, with the increase of studies of chaotic oscillations, the notion of synchronization has been generalized to the latter case. In this context, different phenomena exist which are usually referred to as *synchronization*, so one needs a more precise description to specify them.

Due to a strong interaction of two or a large number of identical chaotic systems, their states can coincide, while the dynamics in time remains chaotic. This case can be denoted as *complete synchronization* of chaotic oscillators. It can be easily generalized to the case of slightly non-identical systems, or the interacting subsystems. A different approach is based on the calculation of the attractor dimension of the whole system and its comparison with the partial dimensions calculated in the phase subspaces formed by the coordinates of each interacting oscillator.

Another well-studied effect is the *chaos destroying synchronization*, when periodic external force acting on a chaotic system destroys chaos and a periodic regime appears (or, in the case of an irregular forcing, the driven system follows the behavior of the force). This effect occurs for a relatively strong forcing as well. A characteristic feature of these phenomena is the existence of a *threshold coupling value* (depending on the Lyapunov exponents of individual systems). In another approach, an *overlapping of power spectra* of certain observables in the interacting systems has been studied. It has been shown that due to interaction, the widths of the peaks of the power spectra may become practically equal, and the peak frequencies become closer. Because of an analogy to the coincidence of frequencies of synchronized periodic systems this effect has been interpreted as synchronization and has been quantified by means of cross-correlation functions and cross-spectra. The effect of phase synchronization of chaotic systems is mostly close to synchronization of periodic oscillations, where only the phase locking is important, while no restriction on the amplitudes is imposed.

Thus, we define phase synchronization of chaotic system as the appearance of a certain relation between the phases of interacting systems (or between the phase of a system and that of an external force), while the amplitudes can remain chaotic and are, in general, non-correlated.

Of course, the very notion of phase and amplitude of chaotic systems is rather non-trivial. Roughly speaking, the phase of an autonomous self-sustained oscillatory system is related to the symmetry with respect to time shifts. Therefore, the phase disturbances do not grow or decay, what corresponds to the zero Lyapunov exponents. The Lyapunov exponent of a dynamical system is a quantity that characterizes the rate of separation of infinitesimally close trajectories. It is common to just refer to the largest one, i.e. to the Maximal Lyapunov exponent (MLE), because it determines the predictability of a dynamical system. If the oscillations are periodic, the phase rotates nearly uniformly, while in the chaotic case the dynamics of the phase is affected by chaotic changes of the amplitude, so one can expect a Brownian (random-walk-like) behavior of the phase.

3.4 Neural Bindings

Synchronization is related to a central issue in neuroscience, neural *bindings* within the same area or between different areas of the CNS. Brain activity is observed at three main space-time scales:

- At the *microscopic scale,* action potentials of single neurons are measured in milliseconds and microns.
- At the *macroscopic scale,* domains of high metabolic demand are imaged in seconds and centimeters (Freeman, 2000, 2002) by a variety of techniques for measuring spatial patterns of cerebral blood flow.
- In between, the *mesoscopic scale* of millimeters (mm) and tenths of a second pattern of massive dendritic potentials can be seen in electroencephalographic (EEG) recordings from waking and sleeping brains.

These waves oscillate in several low frequency ranges, such as theta (3-7 Hz), alpha (8-12 Hz), beta (13-30 Hz), and gamma (30-100 Hz). Long stretches of recordings give broad spectra roughly conforming to $1/f^2$, which have suggested to many researchers the possibility that this activity is governed by chaotic attractors instead of limit cycles embedded in noise. This interesting hypothesis has been pursued with the aid of models derived from deterministic chaotic systems, such as twist-flip maps, Lorenz, Rössler and Chua attractors. These low-dimensional models have engendered highly successful algorithms such as those of Takens, Grassberger, Procaccia and Guckenheimer for estimating the correlation dimensions and Lyapunov exponents of these models. This success has led to efforts for modelling the cognitive functions of sensory cortices in the realm of pattern classification with networks of chaotic elements that are coupled with the nearest neighbor with weighted connections. The proper response of such systems would be synchronization or phase locking of an array of chaotic elements into one of a set of global chaotic attractors, after suitable training to form the attractor landscape.

Attempts to develop and apply such models by calculating correlation dimensions and related measures from EEGs have not succeeded in converging into reliable results. The failure is partly due to an assumption that the stationarity and autonomy of the models holds also true for brain systems, which is inappropriate because *brains are continuously engaged with their bodies and environments, and they undergo repeated state transitions in multiple time scales.* Most importantly, the methods fail because EEGs are heavily mixed with *various kinds of noise*, for some of which the dimensions are infinite. Estimates of correlation dimension invariably increase with the duration of samples.

EEGs, anyway, manifest a small fraction of the total variance of neurons in the generating populations that is covariant (synchronous), and the spatial pattern of phase reveals a critical determinant of the size of the areas of self-organized activity in sensory cortices (~ 1 cm). The group velocity of the state transitions by which spatiotemporal patterns of cortical activity emerge can be successfully extracted from noise by the targets of cortical transmissions (Freeman, 2000).

3.5 Control

Pecora and Carroll suggested that synchronization planning of chaotic systems is possible by introducing appropriate coupling. By doing this, we might succeed in controlling them. They have shown that if a reproduced part (response system or subsystem) of chaotic system responds to a chaotic signal from the full drive system, under some conditions the signal in the response part would converge to the corresponding signals in the drive system. In the following scheme, the drive or master system controls the response or slave system through the replica part (Pecora, 1990, 1991; Carroll, 1993; Abarbanel, 1994).

The afore mentioned procedure is called synchronization of chaotic systems, and further, this idea has been generalized by *cascading* the reproduced parts or subsystems. In this case a chaotic system may be divided into two subsystems, each synchronizing with the full system, when driven by the appropriate chaotic signal.

This cascading effect can eventually produce feedback loops in the master system, as usually happens in living systems, including humans.

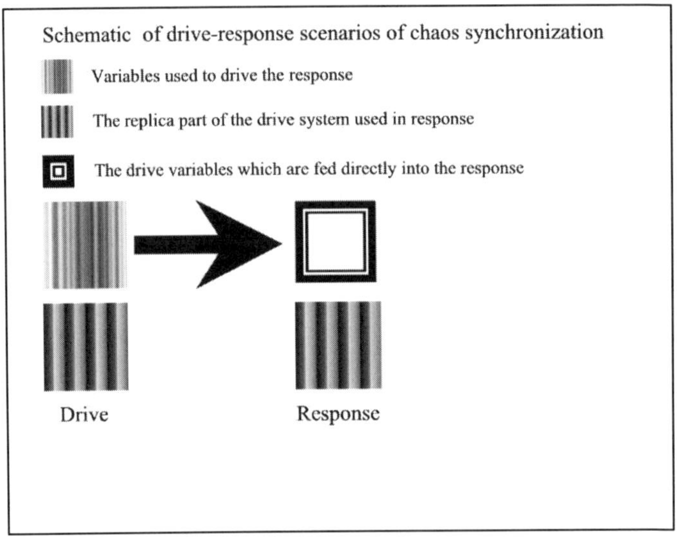
Schematic of drive-response scenarios of chaos synchronization
Variables used to drive the response
The replica part of the drive system used in response
The drive variables which are fed directly into the response
Drive
Response

Fig. 18.

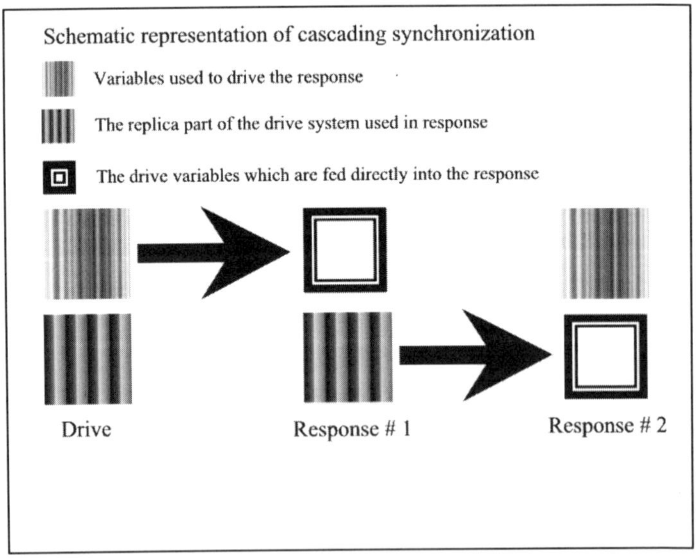
Schematic representation of cascading synchronization
Variables used to drive the response
The replica part of the drive system used in response
The drive variables which are fed directly into the response
Drive
Response # 1
Response # 2

Fig. 19.

This generates a co-evolution of chaotic systems that can be driven by an appropriate selection in the natural, rich, variation of states usual in chaotic systems. That is a matter of controlling their co-evolution.

The control of chaos: exploiting the critical sensitivity to initial conditions to play with chaotic systems. A deterministic system is said to be chaotic whenever its evolution sensitively depends on the initial conditions. This property implies that two trajectories emerging from two different but similar initial conditions will separate exponentially in the course of time.

The fact that some dynamical systems possess such a critical dependence on initial conditions was known since the end of the last century. However, only in the last thirty years, experimental observations have pointed out that chaotic systems are so common in nature. They can be found, for example, in Chemistry, in Nonlinear Optics, in Electronics, in Fluid Dynamics etc. Many natural phenomena can also be characterized as being chaotic. They can be found in meteorology, solar system, heart and brain of living organisms and so on.

Due to their critical dependence on initial conditions and as experimental initial conditions are never known perfectly, these systems are intrinsically unpredictable. Indeed, the prediction trajectory emerging from a *bonafide* initial condition and the real trajectory emerging from the real initial condition diverge exponentially in course of time, so that error in predictions (the distance between prediction and real trajectories) grows exponentially in time, until making the system's real trajectory completely different from the predicted one at long times.

For many years, this feature made chaos undesirable, and most experimentalists considered such characteristic as something to be strongly avoided. Besides their critical sensitivity to initial conditions, chaotic systems exhibit two other important properties. Firstly, there are an infinite number of unstable periodic orbits embedded in the underlying chaotic set. In other words, the skeleton of a chaotic attractor is a collection of an infinite number of periodic orbits, each one being unstable. Secondly, dynamics in a chaotic attractor is periodic, which implies that during its temporal evolution the system periodically visits small neighborhood of every point in each one of the unstable periodic orbits embedded within the chaotic attractor.

A relevant consequence of these properties is that a chaotic dynamics can be seen as shadowing some periodic behavior at a given time, and erratically jumping from one to another periodic orbit. The idea of controlling chaos is when a trajectory approaches every so often, a desired periodic orbit embedded in the attractor, one applies small perturbations to stabilize such an orbit. Switching on the stabilizing perturbations, the trajectory moves to the neighborhood of the desired periodic orbit that can now be stabilized. This fact has suggested the idea that the critical sensitivity of a chaotic system to changes (perturbations) in its initial conditions may be, in fact, very desirable in practical experimental situations. Indeed, if it is true that a small perturbation can give rise to a very large response in the course of time, it is also true that a judicious choice of such a perturbation can direct the trajectory to wherever one wants in the attractor, and to produce a series of desired dynamical states. This is exactly the idea of targeting.

The important point here is: because of chaos, one is able to produce an infinite number of desired dynamical behaviors, either periodic or not periodic, using the same chaotic system, with the only help of tiny perturbations chosen properly. We stress that this is not the case for non-chaotic dynamics, wherein perturbations for producing a desired behavior must, in general, be of the same order of magnitude as the unperturbed evolution of the dynamical variables.

The idea of chaos control was enunciated at the beginning of this decade at the University of Maryland. The ideas for controlling chaos were outlined and a method for stabilizing an unstable periodic orbit was suggested, as a proof of principle. The main idea consisted in waiting for a natural passage of the chaotic orbit close to the desired periodic behavior, and then applying a small judiciously chosen perturbation, in order to stabilize such periodic dynamics (which would be, in fact, unstable for the unperturbed system). Through this mechanism, one can use a given laboratory system for producing an infinite number of different periodic behavior (the infinite number of its unstable periodic orbits), with a great flexibility in switching from one to another behavior. Much more, by constructing appropriate goal dynamics, compatible with the chaotic attractor, an operator may apply small perturbations to produce any kind of desired dynamics, even not periodic, with practical application in the coding process of signals. The perspective of unifying the techniques of deterministic chaos control with a statistical stochastic description as a possible therapeutic strategy against dynamical diseases is surely something to consider (Kupfer et al., 1988).

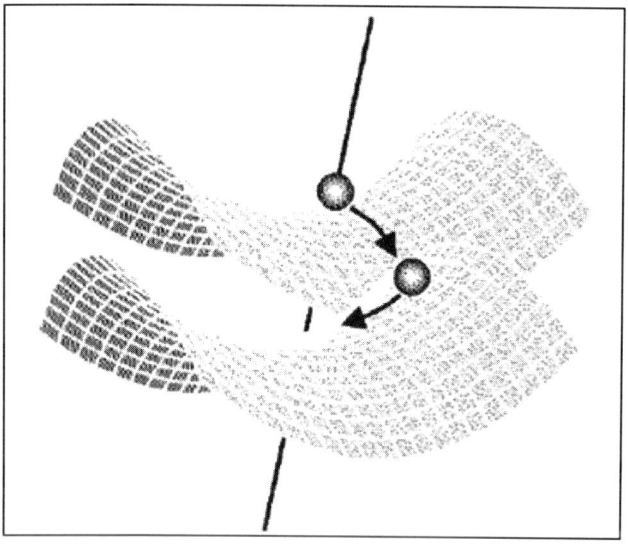

Fig. 20. Local geometry of control: 2D saddle dynamics (Dubè, 2000).

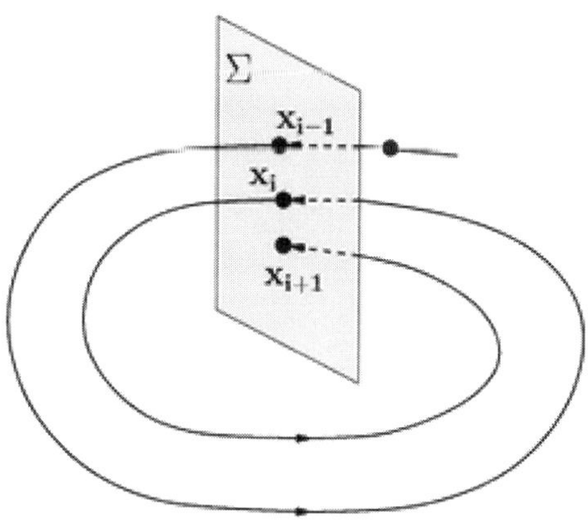

Fig. 21. Poincarè section.

3.6 Co-Evolution

John von Neumann, circa 1950, affirmed: "All stable processes, we shall predict. All unstable processes, we shall control" (Dubé & Desprès, 2000). In his 1985 Giord Lectures, Freeman Dyson expressed his quite different opinion on the matter of chaos:

"A chaotic motion is generally neither predictable nor controllable. It is unpredictable because a small disturbance will produce exponentially growing perturbation of the motion. It is uncontrollable because small disturbances lead only to other chaotic motions and not to any stable and predictable alternative. Von Neumann's mistake was to imagine that every unstable motion could be nudged into a stable motion by small pushes and pulls applied at the right places.

It was not until 1990 however that Ott, Grebogi and Yorke (OGY) addressed the question of control of chaos and described, very much in the spirit of von Neumann, the theoretical steps necessary to achieve this goal (Boccaletti, Grebogi, Lai, Mancini, & Maza, 2000; Dubé et al., 2000). This work was rapidly followed by experimental verification: von Neumann's dream had becoming reality?

It is reasonable to assume that one does not have complete knowledge about the system dynamics since our system is typically complicated and imperfect. It is better, then, to work in the space of solutions since the equations, even if available, are not too useful due to the sensitivity of the dynamics to perturbations. *One gets solutions by obtaining a time series of one dynamically relevant variable.* The *right perturbation* to be applied to the system is selected *after a learning time*, wherein the dependence of the dynamics on some external control is tested experimentally. Such perturbation can affect either a control parameter of the system, or a state variable.

Let us draw the attention on a chaotic dynamics developing onto an attractor in a D-dimensional phase space. One can construct a section of the dynamics such that it is perpendicular to the chaotic flow (it is called *Poincaré section*). This (D-1)-dimensional section retains all the relevant information of the dynamics, which now is seen as a mapping from the present to the next intersection of the flow with the Poincaré section. Any *periodic behavior* is seen here as a periodic cycling among a discrete number of points (the number of points determines the periodicity of the periodic orbit). Since all periodic orbits in the unperturbed dynamics are unstable, also the periodic cycling in the map will be unstable. Furthermore, since, by periodicity, the chaotic flow visits closely all the unstable periodic orbits, this implies that also the mapping in the section will visit closely all possible cycles of points corresponding to a periodic behavior of the system. Let us then consider a given periodic cycle of the map, such as period one. A period one cycle corresponds to a single point in the Poincaré section, which repeats itself indefinitely. Now, *because of the instability of the corresponding orbit, this point in fact possesses a stable manifold and an unstable manifold.* For stable manifold we mean the collection of directions in phase space through which the trajectory geometrically approaches (diverges away from) the point.

The control of chaos idea consists in perturbing a control parameter when the natural trajectory is in a small neighborhood of the desired point, such that the next intersection with the Poincaré section puts the trajectory on the stable manifold. In this case, all divergences are cured, and the successive natural evolution of dynamics, except for nonlinearities and noise, converges to the desired point (that is, it stabilizes the desired periodic behavior). Selection of the perturbation is done by means of a reconstruction from experimental data of the local linear properties of the dynamics around the desired point.

In some practical situations, however, it may be desirable to perform perturbations on a state variable accessible to the operator. This suggests the development of some alternative approaches. The first consists in designing a *proper feedback* line through which a state variable is directly perturbed such as to control a periodic orbit (Pyragas, 1996, 2001).

The second method requires the availability of a state variable for observation and for the perturbations. In such a case, a *negative feedback* line can be designed which is proportional to the difference between the actual value of the state variable, and the value delayed of a time lag T. The idea is that, when T coincides with the period of one unstable periodic orbit of the unperturbed system, the negative feedback pushes to zero the difference between the present and the delayed dynamics, and the periodic orbit is stabilized. Furthermore, as soon as the control becomes effective, this difference goes effectively to zero, so that the feedback perturbation vanishes. Moreover, as before, a *preliminary learning time* is needed, for learning the periods of the unstable periodic orbits. In the above mechanism, the *proportionality constant* entering in the feedback loop is given in where an adaptive technique has been introduced which automatically selects this constant by adaptively exploiting the local dynamics of the system (the Occasional Proportional Feedback OPF method and the Recursive Proportional Feedback RPF method).

3.7 OGY

The idea of Ott, Grebogi & Yorke (1990) (OGY) is that *chaos may indeed be desirable* since it can be controlled by using small perturbation to some accessible parameter or to some dynamical variable of the system.

The major key ingredient for chaos control is the observation that a chaotic set, on which the trajectory of the chaotic process lives, is embedded within it a large number of unstable low-period periodic orbits. In addition, because of periodicity, the trajectory visits or accesses the neighborhood of each one of these periodic orbits. Some of these periodic orbits may correspond to a desired system's performance according to some criterion. The second ingredient is the realization that chaos, while signifying sensitive dependence on small changes to the current state and rendering unpredictable the system state in the long time, also implies that the system's behavior can be altered by using small perturbations. Then, the accessibility of a chaotic systems to many different periodic orbits combined with its sensitivity to small perturbations allows for control and manipulation of a chaotic process. Specifically, the OGY approach is then as follows. One first determines some of the unstable low-period periodic orbits that are embedded in the chaotic set. One then examines the location and the stability of these orbits and chooses one which yields the desired system performance. Finally, one applies small control to stabilize this desired periodic orbit.

However, all this can be done from data by using nonlinear time series analysis for the observation, understanding and control of the system. This is particularly important since chaotic systems are rather complicated and the detailed knowledge of the equations of the process is often unknown.

3.8 Variation And Selection

Targeting desirable states within chaotic attractors is a preliminary of selection and controlling co-evolution. One of the major problems in the above process is that one can switch on the control only when the system is sufficiently close to the desired behavior. This is warranted by the periodicity of chaos regardless of the initial condition chosen for the chaotic evolution, but it may happen that the small neighborhood of a given attractor point (target) may be visited only infrequently, because of the locally small probability function. Thus the unperturbed dynamics may take a long time to approach a given target, resulting in an unacceptably large waiting time for the operator to apply the control of chaos process. Efficient targeting methods can, instead, reduce waiting time, and they can be seen as a preliminary task for chaos control, independent of the particular control algorithm that one applies.

The lessons learned through the previous examples and many more not reported here allow us to draw a list of the properties and advantages of the adopted philosophy on the co-evolution of complex systems and to point to remaining difficulties:

- no model dynamics is required *a priori* and only local information is needed;
- computations at each step are minimal;
- gentle touch is appropriate as required changes can be quite small (<1%);
- multi-purpose flexibility as different periodic orbits can be stabilized for the same system in the same parameter range;
- control can be achieved even with imprecise measurements of eigenvalues and eigenvectors as methods are robust;

- same methods can be applied to synchronization of several different chaotic systems.

At least three complications come to mind when one considers the implementation of chaos control strategies:

- the presence of noise (dynamical or environmental) may induce loss of control or hinder it altogether;
- the average waiting time to fall in the neighborhood may be very (too) long, especially for Hamiltonian systems and a targeting strategy should complement the control method;
- system's parameters may drift with time and this non-stationarity should be accounted for by updating the control information.

Tracking is the name given to this procedure. The drive system has to re-split in an observing/controlling subsystem each time a feedback cascading happens. Cascading can generate feedbacks and lead to co-evolution, though this might also produce uncontrollable drifts in synchronized systems.

However, as the greatest challenge will remain for some time the application to complex biological systems: in particular to mind and brain dynamics, despite early efforts, euphoria has been replaced by pragmatism.

3.9 Many Attractors

Chris Eliasmith (2005) proposed to extend this same approach to networks of many attractors. More precisely, an attractor network is a set of N network nodes connected in such a way that their global dynamics becomes stable in a D dimensional space, where usually N>>D. This assumes that there is no additional external input, and that stability indicates that the network state resides, over time, on some D-manifold (e.g., line, circle, plane, toroid, etc.). It is useful to stress that, if each node is an attractor, the network dynamics could produce a higher order *attractor of attractors*.

In order to avoid confusions, in this *hyperstructure* it is important to keep two domains distinct when discussing attractor networks. These are:

- *network state space* (which is determined by the set of all possible node states); and
- *attractor state space* (which is a subspace of the network state space that only includes points on an attractor).

The network state space is built by different attractors and its own network attractor typology would be similar to lower order attractors: 1) point attractors; 2) line, ring, and plane attractors; 3) cyclic attractors; 4) strange, chaotic attractors.

To have a simplified idea of a network attractor we might consider the energy landscape of a network with multiple attractors, e.g. a Hopfield network (1982). Fixed points are shown as dots. A sample basin of attraction is shown as a dotted circle.

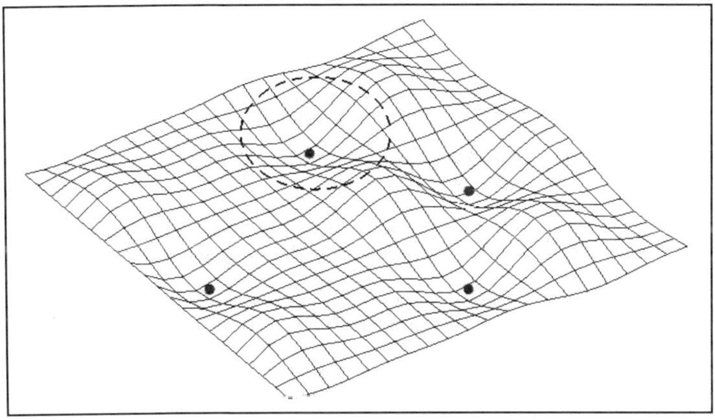

Fig. 22. Local area of an attractor network landscape (Eliasmith, 2005).

Often, these biologically inspired models have adopted *non-biological nodes* (e.g., sigmoid response functions, or rate neurons). However, more recent work has relied largely on spiking models, though of varying degrees of biological plausibility. In addition, there is increasing evidence that many brain areas act as attractor networks (e.g., Wills et al., 2005).

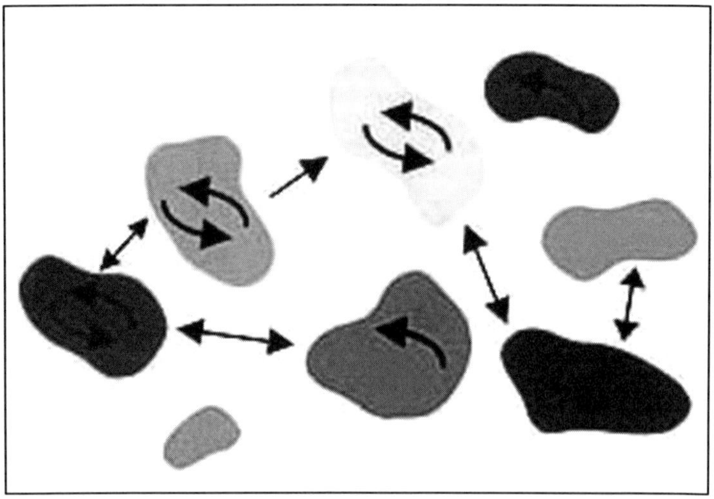

Fig. 23. Transients between different attractors (Eliasmith, 2005).

As the complexity of models continues to increase, attractor networks are likely to form important sub-networks in larger models. This is because the many clear information processing abilities of attractor networks (e.g., categorization, filtering noise, integration, memorization, etc.) make them good candidates for being some of the basic building blocks of large-scale models in living systems.

3.10 Network Theory

The other crucial missing tool in the toolbox we need in order to comprehend Mind Force. is now, obviously, *network theory*. "The study of complex networks is only the next logical step in a larger journey, the quest for a science of spontaneous order" (Strogatz, 1994; Golubitsky, Luss & Strogatz, 1999; Strogatz, 2003). Although each discipline has developed rudimentary models in order to capture relevant networks, one of the most sophisticated of these is the *random network theory* explored in the 1960s by the Hungarian mathematicians Paul Erdős and Alfréd Rényi. But do we seriously believe that real networks are random?

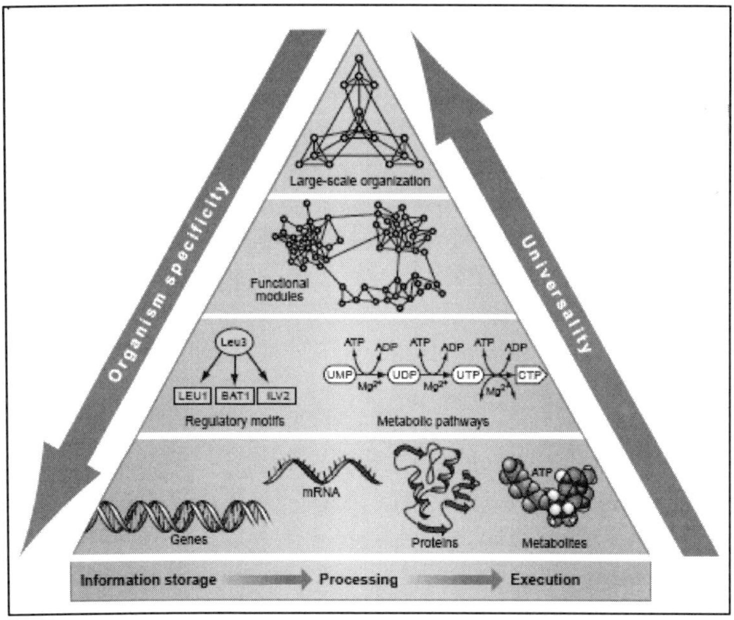

Fig. 24. Network types and distribution (Barabasi, 2005).

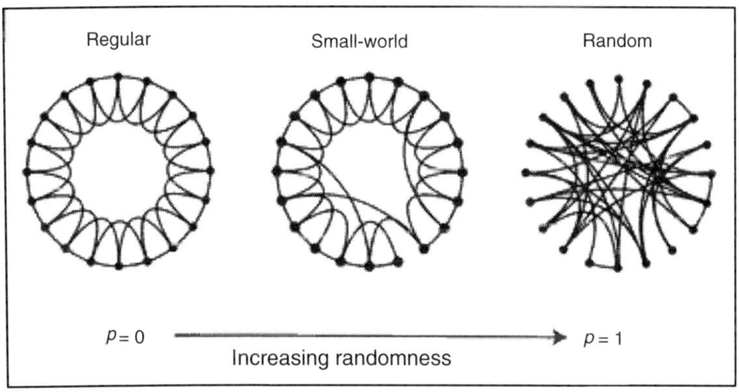

Fig. 25. Network graph representations (Watts & Strogatz, 1998).

Successful functioning of cells or societies must be governed by laws and organizing principles that should be reflected in their architecture (Barabasi, 2005). In referring to network architecture we extend this definition from aggregates in which physical *connections* are spatially stable (for example the brain or the Internet) to networks in which *interactions* can be temporary, *dynamical networks*. Networks of coupled dynamical systems have been used to model biological oscillators, excitable media, neural networks, spatial games, genetic control networks and many other self-organizing systems. One of the most interesting and active research areas in modern network theory is the study of relationships between structure and function, particularly the relation between network topology and synchronization dynamics. Ordinarily, geometry of connections is assumed to be either completely regular or completely random. But, many biological, technological and social networks lie somewhere between these two extremes.

Watts & Strogatz (1998) explored simple network models that can be found in this middle ground: for example regular networks rewired to introduce increasing amounts of disorder. They found that these systems can be highly clustered, like regular lattices, yet have small characteristic path lengths, like random graphs. They called them *small-world networks*, by analogy with the *small-world phenomenon* (popularly known as *six degrees of separation*). Most of dynamical networks in life sciences are small-world networks. The neural networks of animal nervous system, the power grid of nations, and the collaboration graph of film actors are shown to be small-world networks. Models of dynamical systems with small-world coupling display enhanced signal-propagation speed, computational power, and synchronization. The most revealing measure of a network's overall structure is hidden in the *degree distribution*, which is the probability that a node has *k* links (*degree k*) (Barabasi, 2005: 69). In the Internet, the cell or the social network, the degree distribution follows a *power law*, in contrast with the random model. Such networks are *scale-free*.

The *power law* means that real networks are not as "democratic" as the random model suggests, but a few highly connected nodes, or *hubs*, hold together a large number of small nodes. How would we know whom you will meet tomorrow, what page would you link your webpage to, or which molecule will react with an ATP molecule in your nerve cell? The true intellectual thrill for studying complex networks comes from the recognition that despite this microscopic randomness, a few fundamental laws and organizing principles can explain the topological features of such diverse systems as the cell, the Internet or society.

The origin of scale-free topology and hubs in biological networks can be reduced to two basic mechanisms: growth and preferential attachment. *Growth* means that a network emerges through the subsequent addition of new nodes. *Preferential attachment* means that new nodes prefer to link to more connected nodes. Growth and preferential attachment generate hubs through a *rich-gets-richer* mechanism: the more connected a node is, the more likely it is that new nodes will link to it, which allows the highly connected nodes to acquire new links faster than their less connected peers. For example, the probability that one node will connect to another node *a* is twice as large as connecting to node *b*, if the degree of node *a* (*ka = 4*) is twice the degree of node *b (kb = 2)*. In protein interaction networks, scale-free topology seems to have its origin in gene duplication (Barabasi, 2004). Elementary building blocks organize themselves into small recurrent patterns, called *pathways* in metabolism and *motifs* in genetic-regulatory networks.

3.11 Interacting Oscillators

In an interesting study Ernest Lawrence Rossi (1996) tried to draw the whole picture of the psychobiology of mind-body communication as "the complex, self-organizing field of information transduction". Unfortunately he misses most of the contemporary knowledge on synchronization, network theory and field theory, just focusing on information streams and transduction. Anyway, his attempt is interesting, especially for the figure attached at the end of this chapter. This figure, though partial, might give us some hint about the scales, clusters and processes we have been describing as the operations of Mind Force.

It is evident that the resulting hypernetwork is formed by different kind of oscillators: relaxation and non-relaxation; chaotic and non-chaotic; with similar and with quite different frequencies etc.

David Somers & Nancy Kopell (1995) studied waves and synchrony in networks of oscillators of relaxation and non-relaxation type. They showed how chains of phase oscillators that phase-lock produce *travelling waves*. The waves were shown to be a consequence of boundary effects: the oscillators at the ends receive different input than those in the middle, and compensate for the difference by setting up phase differences. They also observed fractured waves and fractured synchrony in networks of relaxation oscillators.

Horacio Zanette has been focusing in this area and in a paper with Alexander Michalkov they studied condensation in globally coupled populations of chaotic dynamical systems (Zanette & Michalkov, 1998). They found that synchronization begins at low coupling intensities with the appearance of small coherent groups of oscillators on the background of the rest of the population performing asynchronous oscillations. The elements belonging to such groups constitute a dynamical condensate. As the *coupling strength* is increased, the number of particles in the condensate grows and eventually the whole population becomes divided into several coherent clusters. Within each cluster, the elements follow the same chaotic dynamical trajectory. Under further increase of the global coupling, the number of coherent clusters gets reduced until full mutual synchronization is achieved. To quantitatively characterize the condensation, they used two order parameters. The first of them is given by the *ratio* r of the number of pairs with zero distances to the total number of pairs. In the absence of a condensate, $r = 0$. On the other hand, $r = 1$ when complete *mutual synchronization* of the whole population takes place. The second parameter s represents the fraction of the population belonging to coherent clusters. It is given by the relative number of elements that have at least one other element with the same state in the considered population. Therefore, s can be viewed as characterizing the size of the condensate.

These forms of synchronization are essential to the functioning of some artificial systems, and have been observed in certain insect populations. On the other hand, they are expected to play a less relevant role in most biological systems, where the complexity of collective functions requires *a delicate balance between coherence and diversity*. Consider, for instance, the brain, where highly coherent activity patterns are only realized under pathological states, such as during epileptic seizures. Many biological systems consisting of interacting agents, ranging from bio-molecular complexes to social populations, are normally found in configurations where the ensemble is segregated into groups with specific functions (Zanette & Mikhailov, 2004).

While the evolution of individual elements is highly correlated inside each group, the collective dynamics of different groups is much more independent. Usually, *clustering* is a dynamical process, where groups may preserve their identity in spite of the fact that single elements are continuously migrating between them. The individual motion towards or away from clusters may also be controlled by the internal state of each element, which favors or inhibits grouping with other elements. This is observed in natural phenomena ranging from complex chemical reactions, where biomolecules react with each other only when they have reached appropriate internal configurations, to social systems, where the appearance of organizational structures requires compatibility between the individual changing states of the involved agents. The behavior of each single oscillator is usually described in terms of differential equations which can be linear or non linear (the latter usually for biological oscillators). Nonlinear dynamics is the study of the complex ways a system evolves over time. The collective behavior of enormous systems of units has been usually described by statistical mechanics. Network theory is a way to connect these complimentary branches of physics. The analytical techniques of statistical physics can be brought to bear the puzzle of how brain cells and other living things manage to synchronize amongst each other (Strogatz, 2003: 55; Albert & Barabasi, 2002). Like light in quantum theory, oscillators can be considered as sine waves or as particles. Networks of oscillators could be modeled as spin glasses, or spin foam. We may even recognize how ensembles of oscillators and networks might present peculiar quantum phenomena like the *Bose-Einstein condensate*, a spontaneous collapse into the same quantum state, the state of lower possible energy (Strogatz, 2003: 134). In this state all the waves are locked in step, they are phase coherent.

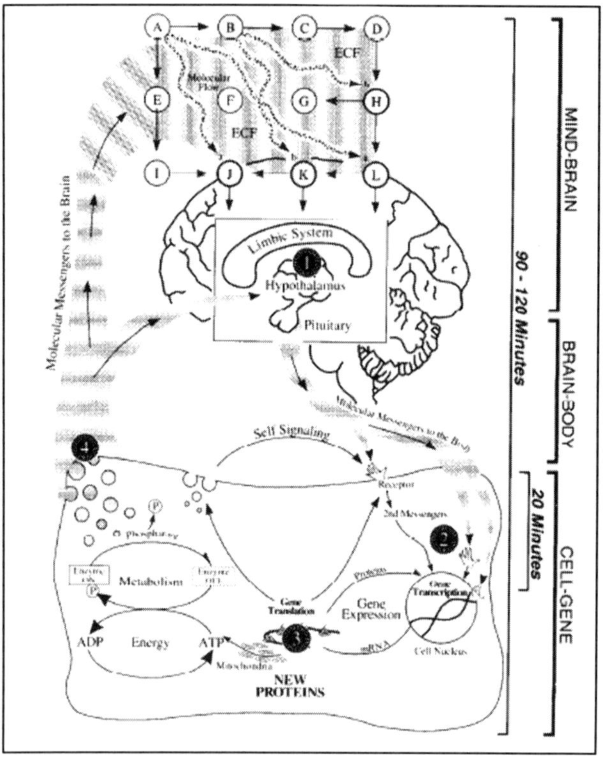

Fig. 26. Heterogeneous couplings in hyperstructure (Rossi, 1996).

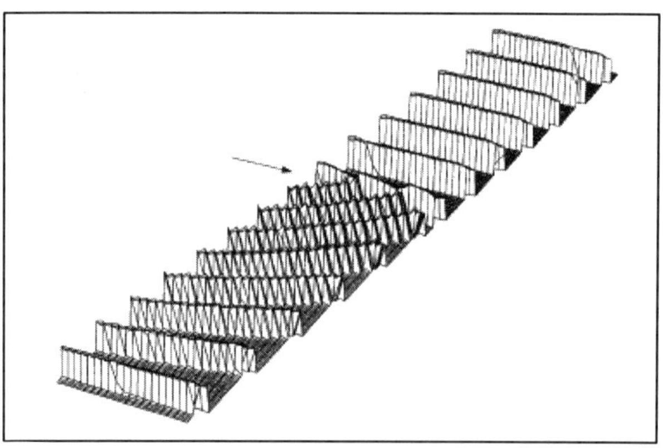

Fig. 27. Start of synchrony and travelling waves in relaxation oscillators (Somers & Kopell, 1995).

Chapter 4

Mind Force

4.1 Matter of Mind

Remarkably little has been written about consciousness in the theory of biological evolution. Richards (1987) captures the core of the problem in his summing up of an argument originally formulated by William James (1879, 1890).

"Consciousness is a manifest trait of higher organisms, most perspicuously of man; like all such traits it must have evolved; yet it could have evolved only if it were naturally selected; but if naturally selected it must have a use; and if it have a use, then it cannot be causally inert. Mind, therefore must be more than an excretion of brain; it must be, at least in some respect an independently effective process that is able to control some central nervous activity" (Richards, 1987: 431). We might add that mind must have a power control over the effectiveness of man on reality.

The idea that an immaterial entity can influence a material entity (reality and the body) is not compatible with an old notion of causality according to which every change in the natural world is produced by contact of spatially extended bodies. This argument was raised by many of Descartes contemporaries. Their primitive antiquated conception of matter as something spatially extended and the related connected notion of causality (as restricted to action by contact) was outpaced by subsequent developments in physics. This conception has not entirely lost its influence in scientific debate and common views. For example P. S. Churchland argues against the existence of "soul stuff" that is not "spatially extended" (1986: 318). Dennett discusses what he calls the *standard objection* which was all too familiar to Descartes (Descartes, 1988; Dennett, 1991) reformulating it in modern terms. In his illustration of this modern criticism, we find the paradox of *Casper the Friendly*

Ghost who is both gliding through walls and grabbing falling towels. These contradictory events seem more of a problem for the adherents of mechanistic notions than to those who are familiar with modern physics.

The *analogy* between mind and forces is not an entirely new concept. In philosophy, Hobbes and Leibniz identified a component of mind which they called *conatus,* with a physical force (Watkins, 1974: 410). In a quite different context, we might mention the case of the so-called *animal magnetism* (Ellenberger, 1970). The term's most common usage today refers to a person's sexual attractiveness or raw charisma. Yet, Animal Magnetism (in French *magnétisme animal*) originally signified a magnetic fluid or ethereal medium residing in the bodies of animate beings, as postulated by Franz Mesmer. The term translates Mesmer's *magnétisme animal.* Mesmer chose the word 'animal' to distinguish his supposed vital magnetic force from those referred to at that time as mineral, cosmic and planetary magnetism.

The existence of Mesmer's magnetic fluid was scientifically examined by a French Royal Commission set up by Louis XVI in 1784. Whilst the Commission agreed that the treatments claimed by Mesmer were indeed effective, the Commission also concluded there was no evidence of the existence of his magnetic fluid, and that its effects derived from either the imaginations of its subjects or through charlatanry. In those same years, Abbé Faria introduced oriental hypnosis to Paris. He was the first to cause a breach in the theory of magnetic fluid, by placing focus on the importance of suggestion, and to demonstrate the existence of autosuggestion.

Mesmerism, named after Mesmer himself, and hypnosis (as the term is now understood) have nothing in common except their shared historical roots. Furthermore, the experience of the mesmerized subject is significantly different from that of the hypnotized subject.

4.2 Fields

We have already seen how, in different ways, some of the key founders of modern psychology, such as Sigmund Freud, William James and Carl Jung, presented issues in favor of Mind Force (MF). The history of this construct progressed during the following years.

During the late 40s, a social psychologist, Kurt Lewin (1890-1947) born in Germany, immigrated to the USA because of World War II. He established also the Research Center for Group Dynamics at Massachusetts Institute of Technology. Lewin stated that: 'One should view the present situation — the *status* — as being maintained by certain conditions or forces' (Lewin, 1943a: 172). Lewin, who was well known for using concepts like *life space* and *field theory*, proposed to view the social environment as a dynamic field affecting human consciousness. In turn, the person's psychological state influences the social field or milieu.

Lewin sought to describe group life, and to investigate conditions and forces, which bring about change or resist change in groups. In his field theory, a *field* is defined as 'the totality of coexisting facts which are conceived of as mutually interdependent' (Lewin, 1951: 240). Lewin believed that in order for change to take place, the total situation has to be taken into account. If only part of the situation is considered, a misrepresentation of the picture is likely to develop. The whole psychological field, or *lifespace*, within which people acted, had to be viewed, in order to understand human behavior. Within this framework, individuals and groups could be seen in topological terms (using map-like representations). Individuals participate in a series of life spaces, (such as the family, work, school and church) and these spaces were constructed under the influence of various force vectors (Lewin, 1952). His active approach could be summarized in his motto: "Learning is more effective when it is an active rather than a passive process: if you want to truly understand something, try to change it." (Lewin, 1952).

Harry S. Sullivan (1892–1949), like Lewin, was concerned about using theoretical constructs featuring falsifiable reference in interpersonal behavior. While Sullivan may not have read Lewin, Wertheimer, Kohler, or Koffka, he almost certainly would have had welcomed their general approach, since he was proposing a psychological field theory similar to their psycho-physical field theory. He may have seriously intended only a loose heuristic function in casting interpersonal relations as occurring in a "field." Nevertheless, he was intrigued by this notion. It seemed to him that the field provided an accurate and useful description of effects. Though he didn't write much about fields and mind force, during his brief lifespan, he left a wonderful sketch presented during one of his last lectures (LaForge, 2004).

4.3 Force

More recently, the philosopher Karl Popper has emphasized the similarities between mind and forces: "Minds are located, unextended, incorporeal, capable of acting on bodies, dependent on body and capable of being influenced by bodies. (…) Now, I say, things of this kind *do* exist, and we all know it. So, what are these things? These things are forces" (Popper, 1993: 168).

Popper seems to go further than a mere analogy, and he proposes, as a hypothesis "that the complicated electro-magnetic wave fields which, as we know, are part of the physiology of our brains, represent the unconscious part of our minds, and that the conscious mind — our conscious mental intensities, our conscious experiences — are capable of interacting with this unconscious physical force fields, especially when problems *need* to be solved. That *need* is what we call *attention*" (Popper, 1993: 179).

Here Popper is considering unconscious as synonymous with the physical force fields. In the figure in the following page, you might notice that there is an area of mind/brain overlap. This area represents a form of biophysical unconscious. Popper points out that conscious mind may "sink into physiology" and become unconscious: "a mergent process, a process where (unconscious) mind and brain are no longer distinguishable" (Popper, 1993: 171).

In explaining part of his conception of unconscious, Popper utilized the example of learning a complex psychomotor skill (playing the piano, grasping a mug, and a lot of other activities). In the first stages of the learning process we are conscious and our attention is focused on each single step in the new skills to learn. This stage, sooner or later, disappears and we no longer think about each single step. The new

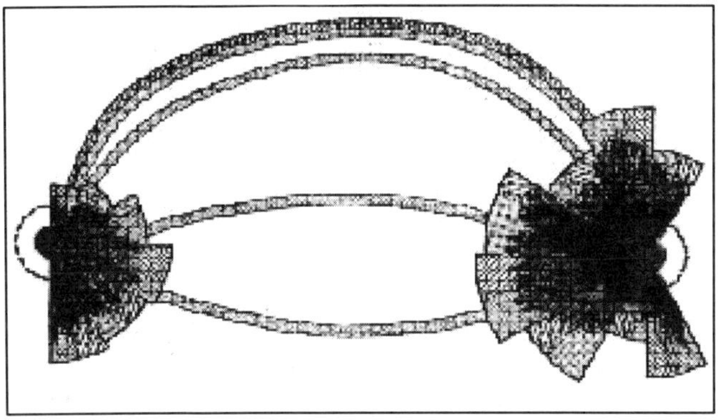

Fig. 28. H. S. Sullivan sketch on mind force fields in a couple (LaForge, 2004).

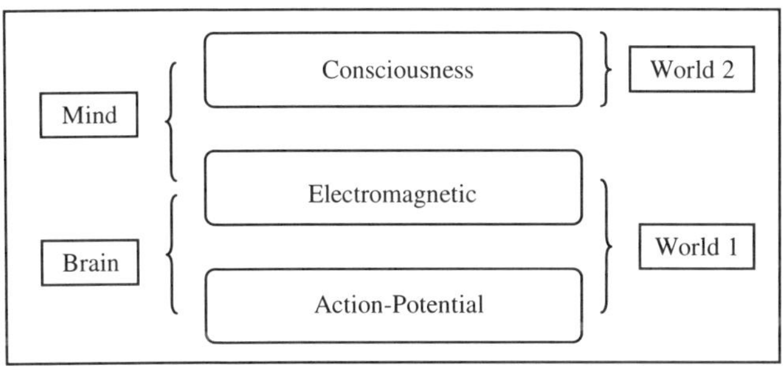

Fig. 29. K. Popper's scheme.

acquired skills are now embodied and, we might say, merged in our being. This process is part of the constitution of *procedural memory*.

There is no need to postulate a mind force in order to explain procedural memory, so Popper's example might be misleading. His proposal of founding mind force on electromagnetic interactions is also improbable for physical reasons. Anyway, some of Popper remarks and his endorsement of the necessity of MF theory are certainly important.

4.4 Conscious Mental Fields

Benjamin Libet, in the same years, had also proposed the hypothetical existence of a *Conscious Mental Field* (CMF) (Libet, 1993b, 1994). The

CMF would emerge as a function of neural activities in the brain and it would have the attribute of a conscious subjective experience. Libet also suggested that it could act back on certain neural activities and would therefore affect behavior, as in a willed action. It would account for the unity of a subjective experience, even though the latter emerges from the myriad of activities of billions of nerve cells and their synaptic and non-synaptic interplays.

In Libet's opinion "The CMF, like the subjective experiences constituted in it, would be accessible only to the individual having the experiences it could not be directly observed by any external physical device except indirectly, by any effects it introduces on behavioral outcomes (just as conscious will is evidenced)" (Libet, 1996: 223).

In a paper entitled "Mind as a force field: a new interactionistic hypothesis", B. Lindahl and P. Arhem (1994) proposed a critical review of some modern approaches to Mind Force. Libet in his response to their remarks (1996) wrote that he liked "Popper's idea of viewing the mind as a kind of force field". Such a CMF force would have been different from all known physical forces, though Popper's hypothesis does not appear to spell out any attributes of that conscious physical force field except in its ability to interact with another entity. The electromagnetic field representing the unconscious mental functions is "doubtful based on evidence available". Consciousness, following Libet's experiments, can simply be a function of the duration of cerebral activations to achieve awareness. Libet was stressing that, "Evidence suggests that conscious functions involve some special neural activities that are simply added to those involved in conscious functions" (1996: 224). Libet is posing two important questions: first, how does the CMF arise out of mental activities; and second, exactly how the CMF does act on the physical brain.

Nevertheless, it seems that Libet was forgetting the dynamical nature of CMF and the self-organizing nature of its power. He does revisit the issue, reminding, "whether electromagnetic fields are representative of unconscious mental functions could be tested in principle by experimentally distorting and/or disrupting or modifying such fields in the putative relation to unconscious functions". Sperry (1947) had already tried such experiment by cutting the monkey cortex in slices.

However, Sperry's vertical cuts in the cortex may not have affected larger field currents as electrical pathways over and below the cuts were still present as a potential role for over-arching electric fields therefore remains possible.

4.5 The Aconscious

Lindahl & Arhem (1996) noted that "in the mental force field hypothesis only the unconscious part of mind is explicitly interpreted in field terms". They correctly point out that Libet (1996) mistakenly assumed that the Popper force field hypothesis was applied to the conscious mind. Popper left open the question of the nature of conscious mind However, he did except to say that the mind in its general form emerges from the body, somehow, but is not reducible to it. What Libet misunderstood, and it might be not clear enough in Lindahl & Arhem response, is that the unconscious in question should be properly named *aconscious*. Its contents and processes are not repressed, and possibly recalled to consciousness, but rather embedded in the human psychomotor structure via psycho-physiological storage.

Libet, more recently (2006), has summarized his view on Mind Force. He starts by recognizing that, if there is a generally held assumption that mind and brain can interact, this indicates that "two phenomenological entities exist". After this "back to Descartes" statement, he then continues with the questionable assumption that mind, in his view, is just a subjective experience, accessible only by individual introspection.

This point of view, discards all the psychological and linguistic research studying psychomotor, perceptual and linguistic manifestations or derivatives of mind functioning. His point recalls the "private language" paradox proposed by Ludwig Wittgenstein (1967). The conclusion raised by this paradox is that every form of language and thinking (even the so-called inner dialogue) relies on our shared forms of communication. Language (also in the form of culture) is pre-existing to the subject, though its communicative or private usage for thinking can be personalized. If the idea of a private language is incoherent, then it would follow that all language is essentially public: that language is at its core a social phenomenon. For instance, if one cannot have a private language, it might not make any sense to talk of private sensations such as *qualia*; nor might it make sense to talk of a word as referring to a concept, if a concept is understood to be a private mental representation.

Libet recalls the interest that Sir John Eccles derived from Sir Charles Sherrington for the interactions between brain and mind. When the physicist, Henry Margenau, provided a view of the mind as a field that could interact with the brain even with no energy expenditure (Margenau, 1984), this supported Eccles' bias on the nature of mind–brain interaction.

Libet notes, "It is especially noteworthy that Eccles' models of mind–brain interaction were presented without any experimental evidence or experimental designs for testing. That was due at least partly to the untestability of the models. Curiously, an absence of experimental testability did not bother Eccles. When asked if his view that a field of ''psychons'' (his units of mental function; see Wiesendanger, 2006) could mediate the unity of subjective experience (Eccles, 1990) was untestable, Eccles replied that he knew of no way to test that hypothesis (personal communication). Nevertheless, he argued that the hypothesis had explanatory power, and, as such, he believed it had some usefulness and even validity. Eccles produced a stimulus further contribution to work in the direction of an MF definition, but his models remained untested, and, apparently, he was not bothered about it.

Libet also raises the testability issue about the approach proposed by Hiromi Umezawa and his followers: as they proposed a mental field model, which they termed a *Quantum Field Theory* (claimed by this group of authors as different from *Quantum Mechanics*) (Ricciardi & Umezawa, 1967; Umezawa & Vitiello, 1986; Vitiello, 1995, 2001, 2002). In Libet's opinion, "Their model is mostly mathematical, however, and it is not clear how it can be tested." (Libet, 1993; Libet, Freeman, & Sutherland, 1999). Libet is raising a similar objection against the quantum mechanics approach as in the interpretation of quantum theory by Neils Bohr (1885–1962) mind and matter are two aspects of one undivided process. David Böhm (1917–1992) adopted this idea (see Böhm and Factor, 1985). However, this does not solve the problem of how the neuronal activity aspect relates to the subjective, non-physical aspect of mind. If subjective experience is a non-physical phenomenon, what is it? The merit of the Bohr-Böhm approach is in recognizing that there is a physical process behind and beyond mind and brain.

Libet claims that his CMF theory is potentially testable as he described a design for conducting such tests. The proposed experimental test is simple in principle but difficult to carry out, a small slab of sensory cortex, which keeps the tiny cortex island alive by preserving the blood vessels providing blood flow from the arterial branches that dip vertically into the cortex. "The prediction is that electrical stimulation of the sensory slab will produce a subjective response reportable by the subject. That is, activity in the isolated slab can contribute by producing its own portion of the CMF." (Libet, 2006: 324). He states, and we agree with him, that his CMF is an emergent and localizable system property. Less clear is why this experiment, which doesn't seem technically so difficult, has not been performed yet by other scientists or Libet himself.

Libet is also referring to the functioning of the CMF as the delay in sensory awareness of 0.5 s after the initial response of the cortex as well as the other very interesting phenomenon related to readiness for action, which is preceding actual actions by about 300 ms. So, both perceptual and motor activities have significant delays with consciousness. *These empirical findings support the autonomy of aconscious Mind Force processes from consciousness processes.* This landmark, (though still partially controversial) findings by Libet, could entirely re-design the role of conscious and unconscious processes. The term unconscious here can be mistakenly confused with the traditional (repressed) unconscious, but in this case, we are dealing with events without any conscious representation. Indeed, most mental events are unconscious, or we might better say, *aconscious* (Orsucci, 2002b, 2002a). Libet concludes his review mentioning *in vivo* and brain imaging research supporting his findings. "If an experimental test of the CMF was to be carried out, like that described above, it might confirm or contradict the kind of alternatives possible for a mind–brain interaction" (Libet, 2006: 326).

4.6 Energy

Walter J Freeman, after his explorations on mass action in the nervous system, chaos dynamics in perception and even social dynamics, has more recently come to propose a Mind Force approach (Freeman, 2007). He first defines the framework of this approach: "Consciousness fully supervenes when the 1.5 kgm mass of protoplasm in the head directs the body into material and social environments and engages in reciprocity. While consciousness is not susceptible to direct measurement, a limited form exercised in animals and pre-lingual children can be measured indirectly with biological assays of arousal, intention and attention." It is a remarkably non-deterministic and interactionistic approach.

After a general description of the multiple levels of interactions involved, from the molecular to the social levels, including their intermingling, he states: "Every reflex and intentional act and thought is based on the exchanges of matter and energy through neural activity at every scale." (Freeman, 2007: 1022). There is a need for a universal language to comprehend all the incredible mesh of interactions involved and the mathematical tools needed might be already at hand.

There is no definition of what consciousness is and, no physiological or cognitive index of consciousness, as many discussions on consciousness still tend to confuse it with self-consciousness. So, Freeman states that he wants to consider the perceptual and behavioral derivates of consciousness that we might find even in infants and animals, "I leave the hard problem (Chalmers, 1995) to philosophers."

He stresses that the stream of consciousness is cinematographic, as we have seen in Chapter 2 rather than continuous. Consciousness role in human behavior is judgmental rather than enactive, so that its prime role is not to make decisions but to delay and defer action and thereby minimize premature commitment of limited resources. Just as we use to say in the adage, "stop and think before acting". Following this path, Freeman comes to a clear statement: "consciousness is not merely 'like' a force; it is a field of force that can be understood in the same ways that we understand all other fields of force (Freeman, 2004) within which we, through our bodies, are immersed, and which we, through our bodies, comprehend in accordance with the known laws of physics." (Freeman, 2007: 1022).

The models that Freeman has implemented are schematized in two ways: one is the so called Katchalsky model (or K-sets) (Freeman, 1975; Kozma & Freeman, 2003); the other one is the quantum field model he more recently developed in collaboration with Giuseppe Vitiello (2006).

In Freeman's description Mind Force might be found in the action–perception cycle as described by Piaget (1930) and Merleau-Ponty (1942). The cycle begins with a macroscopic state in the brain that embodies a goal. It emerges in the brain from a predictive state implicitly containing nested mesoscopic activity patterns, constructed in corticostriatal and corticocerebellar modules (Houk, 2005). The predictive expectations embedded in sensory cortices are described as landscapes of chaotic attractors within the brain state space. The dynamic memory embodied in nerve cell assemblies is manifested in spatial pattern of amplitude modulation, mostly in the gamma band range. An interesting property of the system is that these dynamical landscapes lack invariance, as they change whether the same stimulus is reinforced or not, or the context is different or the sequence of stimuli is different.

"The attractor governs the neural interactions that generate an oscillatory field of neural activity called a wave packet" (Freeman, 1975). Fields are not fixed representations of the stimuli, and stimuli are not grounded in any fixed way. Each action-perception frame is separated from the others by phase transitions. Freeman cites Wolfgang Köhler who was (1940: 55) quite explicit about this: "Our present knowledge of human perception leaves no doubt as to the general form of any theory which is to do justice to such knowledge: *a theory of perception must be a field theory*. By this we mean that the neural functions and processes with which the perceptual facts are associated in each case are located in a continuous medium". Regrettably, Köhler identified his perceptual field with the epiphenomenal electric field of the EEG, of which the Coulomb forces are much too weak to synchronize the observed oscillations in wave packets (Freeman & Baird, 1989). Sperry (1980) and Pribram (1971) easily disproved this subsidiary hypothesis, with the unfortunate outcome that mainstream neuroscientists largely abandoned field hypotheses.

We are honored of a collaboration and friendship with Walter Freeman, lasting for the last 15 years. During our frequent discussions and correspondence we shared similar views on complexity models, though we come from different background and professional experiences. Recently we came almost at the same time and independently, through sharing parts of the same "incubator", to a radical formulation of the Mind Force construct. The positions expressed in this book are, in any case, exclusive responsibility of the author.

Complexity theory has provided the empirical and mathematical tools to prove that the brain patterns correlated to the cine-like frames in the action-perception cycle are like bubbles in a pan of boiling water at the critical temperature. They can be seen also as the avalanches on a sand pile, as in the model of *self-organized criticality* proposed by Per Bak (Bak et al., 1978; Bak, 1990, 1996).

From this neurodynamical point of view, the personal identity is "embodied in the entirety of the brain-body dynamics" this is the reason why we have been speaking about a comprehensive *biophysical identity*. We find this definition more grounded and explicit, but in the same line of the approach suggested, with different nuances, as *embodiment* (Varela, Thompson & Rosch, 1991), *proto-self* (Damasio, 1999), *global workspace* (Baars, 1999), *synaptic self* (LeDoux, 2002).

4.7 Four Pillars

Following the history of the Mind Force construct we might realize how it comes gradually to take the shape of a new theory, approaching the stage in which it could be formalized. We can summarize some of the necessary prerequisites of this theory.

In order to recognize the existence and operability of MF and its related phenomena, we need to accept and stabilize a definitive transcending of the notorious Cartesian dichotomy. The current discussion on Descartes' error is often missing two important points: the heuristic value that his position has had for centuries in the advancement of science; and a full recognition of all the implications that discarding his approach will have on our new scientific approaches. Not to mention that many crypto and non crypto-Cartesians are still at work. MF is beyond *Res Cogitans* and *Res Extensa* in dynamical and structural terms. We might say that it constitutes a superior dynamical unity.

New physics and new biomedicine gave us some crucial tools to transcend Descartes. The immense complexity and dimensionality of human systems, if considered in a post-Cartesian view, must be studied in the modern terms of complexity theory, nonlinear dynamics, field theory, quantum mechanics, molecular biology and cognitive science. These seemingly different approaches would be integrated in order to reach a real view of MF nature and operations, beyond the dichotomies and appearances we are used to see.

A logical consequence is that MF, in its structure (that we are going to recognize in networks) and dynamics (that we are going to recognize in fields/waves of synchronizing nonlinear oscillators), would be heterogeneous. Dynamics, fields and the hyperstructure of MF would span through molecular domains, neural domains, cognitive domains and even socio-cultural domains. We need to consider how MF fields "pack" specific *synchronous dynamics* "vertically" ranging across these different domains, just we have considered *diachronic dynamical fields* spanning "horizontally" within a single domain.

If we are able to accomplish this reframing of our perceptual and cognitive habits in order to recognize Mind Force, we might see how it forms a dynamical "glue" ensuring attractions in bodies, minds and social ensembles and the cohesion of our inner and outer bio-psychosocial realities.

A definition of Mind Force would be as the dynamical hyperstructure formed by networks of synchronized oscillators coupled in fields spanning through heterogeneous domains (Orsucci, 2009).

In conclusion, Mind Force theory is based on the integration of 4 main pillar-theories:

- Complexity theory
- Synchronization theory
- Network theory
- Field theory

Within the population of bio-psychosocial oscillators some would act as a master hub, others as a slave or a free (Nicolis & Prigogine, 1977; Kauffman, 1993, 1995, 2000; Kelso, 1995) node within the hyperstructure dynamics. Waves of massive and eventually heterogeneous transient entrainment would form attractors and fields. These waves of massive synchronizations propagate through different media, domains and scales. A logical consequence is that MF transient or steady fields would interfere and interact, producing an MF resultant, forming MF dynamical landscapes.

In conclusion, to use an expression coined by Douglas Hofstadter (2007: 39) Mind Force is the result of "the causal potency of collective phenomena" and patterns.

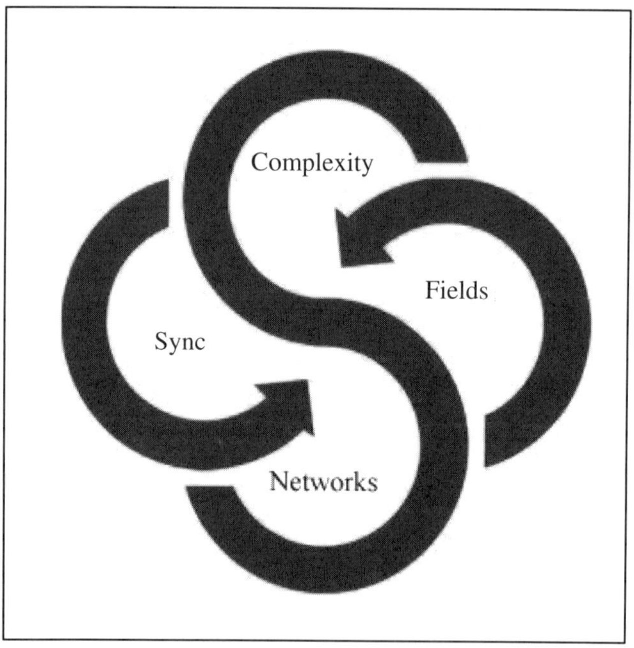

Fig. 30. Four pillars of Mind Force.

Chapter 5

Flows

5.1 A Space Odyssey

Since the first production of tools from the beginning of human presence on earth, evolutionary leaps mark the age of human development. In the movie *2001: A Space Odyssey* (Ambrose, 2001), a savannah-dwelling ape has a eureka-like flash of inspiration in realizing that the tremendous power of the bone tool in his hands. He tosses it skyward, where it morphs into a space station at the dawn of this millennium. Implications such as these specialized forms of symbiotic embodiment with tools in environments represent one of the main factors for human evolutionary processes.

A study of reflexing interfaces might shed more light on Mind Force phenomenology and its flow throughout different psychosocial domains. The cognitive neuroscience of the reflexive function can be one of the main keys to understand how the emergence of new interfaces yields new ways of projecting the human presence and consciousness in the world, one of the expressions of Mind Force. In recent times, Information Science and Technology are accumulating ground for new possible evolutionary leaps. Computing devices, molecular biology, and new media (all members in different ways of the IST set) are redesigning the human embodiment and its environment. An integrated approach of IST and neuroscience can design a map for new possible human evolutions.

Stone tool technology, robust australopithecines, and the Homo genus appeared almost simultaneously 2.5 Million years ago. Once this adaptive threshold was crossed, technological evolution continued to be associated with increased brain size, population size, and geographical range. Traits of behavior, economy, mental capacities, neurological functions, the origin of grammatical language, and socio-symbolic systems have been inferred from the archaeological record of Paleolithic technology (Ambrose, 2001). *Homo Habilis* is, obviously, considered the first toolmaker. The contiguity in the brain of *Broca's area*, linked oral — facial fine motor control and language, to the area for precise hand motor control might be more than casual. The hand of *Homo Habilis* resembles that of modern humans. Its brain was significantly larger (600 to 800 cm^3) than that of earlier and contemporary australopithecines and existent African apes (450 to 500 cm^3). Its teeth were relatively small for its body size, suggesting a relation between tool use, quality of diet, and intelligence.

The production of tools and artifacts is also linked to the development of language, culture and cognitive functions. This happened as tools and artifacts were, just as other socio-linguistic processes, *mediating and reflexing interfaces* in environmental and social interactions. We have yet to discover more about the ways in which speaking, tool-using, and interaction are interwoven into the texture of daily life of contemporary human groups. The birth of *technique* was incubated in the complex system of material resources, tools, operational sequences and skills, verbal and nonverbal knowledge, and specific modes of work coordination that come into play in the fabrication of material artifacts. It is a process, a complex *interplay* of *reflexivity* between sensory-motor skills, symbolic cognition, tools, artifacts and environment. It is a cascading of couplings which can be comprehended within the biophysics of Mind Force.

5.2 Affordances

James J. Gibson (1979), in this context, originally proposed the concept of *affordance* to refer to "all action possibilities" latent in a specific environment, objectively measurable, and independent of the individual ability to recognize those possibilities. Furthermore, those action possibilities are dependent on the physical capabilities of the subject. For instance, a set of steps with risers four feet high does not afford the act of climbing, if the actor is a crawling infant. Therefore, we should measure affordances along with the relevant actors.

Donald Norman (1988) introduced the term *affordance* in Human Machine Interaction, which made it a very popular. Later Norman (1999) clarified that he was actually referring to a perceived *affordance*, as opposed to an objective *affordance*. This new definition illustrated that *affordances* are determined not only by the physical capabilities of the subject, but also by the individual and social knowledge embedded in objects and interactions of everyday life.

For example, if a person steps into a room with a mug and a book, Gibson's definition of affordance allows for a possibility that the subject in question may look at the chair and sit on the book, as this is objectively possible. Norman's definition of perceived affordance however, captures the likelihood that the person will sit on the chair and look at the book, because of the embodiment and social knowledge embedded as clues in these objects.

The significance of evolutionary theory to human sciences cannot be fully appreciated without a better understanding of how phenotypes in general, and human beings in particular, modify significant sources of selection in their environments, thereby co-directing subsequent biological evolution. Empirical data and theoretical arguments suggest that human techno-cultural activities have influenced human genetic evolution by modifying sources of natural selection and altering genotype frequencies in some human populations (Bodmer & Cavalli-Sforza, 1976). Techno-cultural traits, such as the use of tools, weapons, fire, cooking, symbols, language, and trade, may have also played important roles in driving hominid evolution in general and the evolution of the human brain in particular (Aiello & Wheeler, 1995). It is more than likely that some cultural and scientific practices in contemporary human societies are still affecting human genetic evolution. Modern molecular biologists do interfere with genes directly on the basis of their acquired scientific experiences, though this practice might be too recent to have already had an enduring impact on human genetic evolution. Nevertheless, it already brings a new reflexive loop in our development.

Other evolutionary biologists maintain that culture frequently influences the evolutionary process, and some have begun to develop mathematical and conceptual models of gene-culture coevolution that

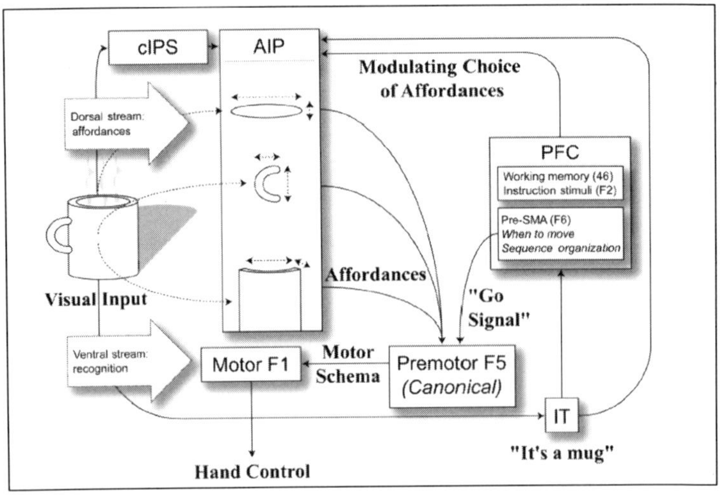

Fig. 31. Neuro-cognitive dynamics in affordance, for example, grasping a mug (Arbib, 2001; Arbib & Mundhenk, 2005).

involve descriptions not only of how human genetic evolution affects culture but also of how human culture can drive, or co-direct, some genetic changes in human populations (Feldman & Laland, 1996). These models include culturally biased non-random mating systems, the treatment of human sociocultural or linguistic environments as sources of natural selection (Aoki & Feldman, 1987), and the impact of different cultural activities on the transmission of certain diseases (Durham, 1991).

5.3 Niche Construction

The common element among all of these cases is that *cultural processes change the human selective environment, thereby affecting which genotypes survive and reproduce.* Culture functions on the basis of various kinds of transmission systems (Boyd & Richerson, 1985), which collectively provide humans with a second, non-genetic "knowledge-carrying" inheritance system. *Niche construction* from all ontogenetic processes modifies human selective environments, generating a legacy of

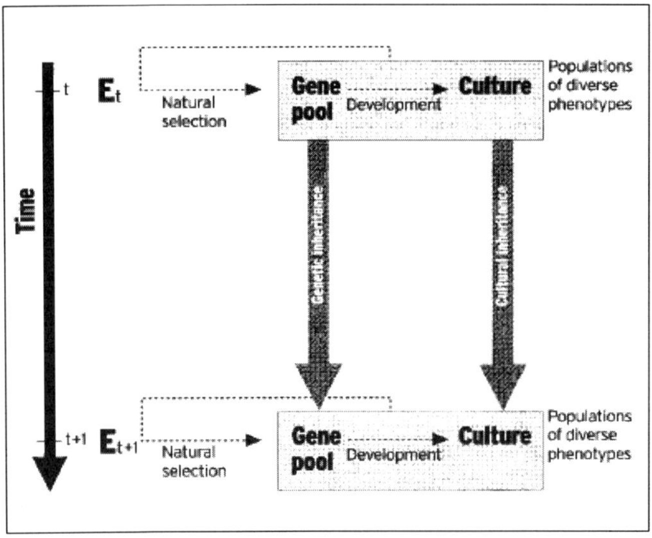

Fig. 32. Evolutionary dynamics involving genes and techno-culture (Laland et al., 2000).

modified natural selection pressures that are bequeathed by human ancestors to their descendants. This figure best captures the causal logic underlying the relationship between biological evolution and cultural change (Laland et al., 2000).

If the techno-cultural inheritance of an environment-modifying human activity persists for enough generations to produce a stable selective pressure, it will be able to co-direct human genetic evolution. For example, the culturally inherited traditions of pastoralism provide a case in point. Apparently, the persistent domestication of cattle, and sheep along with associated dairy activities, indeed altered the selective environments of some human populations for sufficient generations to select for genes that today confer greater adult lactose tolerance (Durham, 1991). Although other animal species have their own proto-cultures (Galef, 1988), it has generally been assumed that Homo sapiens is the only existent species with a techno-cultural transmission stable enough to co-direct genetic evolution (Boyd & Richerson, 1985). We may conclude that our techno-culture is part of our ecological niche.

Building on ideas initially developed by Lewontin (1983), Laland has previously proposed that biological evolution depends not only on natural selection and genetic inheritance but also on *niche construction* (Laland et al., 1996a). Niche construction refers to the activities, choices, and metabolic processes of organisms, through which they define, choose, modify, and partly create their own niches.

For example, to varying degrees, organisms choose their own habitats, mates, and resources to construct important components of their local environments such as nests, holes, burrows, paths, webs, dams, and chemical environments. Many organisms also partly destroy their habitats, through stripping them of valuable resources or building up detritus, processes we refer to as negative niche construction.

One theoretical interpretation that captures some, but not all, of the consequences of niche construction is Dawkins's (1982) *extended phenotype*. Dawkins argues that genes can express themselves outside of the bodies of organisms that carry them. For example, the beaver's dam is an extended phenotypic effect of beaver genes. Like any other aspect of the phenotype, extended phenotypes play an evolutionary role by influencing the chances that the genes responsible for the extended phenotypic trait will be passed on to the next generation. Dawkins emphasizes this single aspect of the evolutionary feedback from niche construction.

An example of contemporary environmental niches can be found in the computer mouse and its related iconic desktop-like interface. Every human interface tends to use biomechanical and physiological properties of the human body in order to reach a possible *perfect symbiosis* between man and machine.

The result is a possibility of a real time interaction with a real or a conceptual object within a learning environment based on augmented reality adding new dimensions to our usual everyday reality. Meanwhile, it also gives at the same time, a new reality to scientific mental objects. Similar, yet already commercial pointing devices, are the Wii Remote for videogames, controllers as inviting as sophisticated, fusing the familiarity of a remote control with motion or neurophysiologic sensing technology. Other experimental devices or prototypes might be interesting examples: eye movements, and brain waves and other bio-signals are captured and amplified to translate bio-signals into useful logic commands and neural-signal interpretation, such as emotions.

5.4 Mirrors

In *Language within our grasp,* Rizzolatti & Arbib (1998) showed that the mirror system in a monkey is the homologue of *Broca's area*, a crucial speech area in humans, and argued that this observation provides a neurobiological "missing link" for the long-argued hypothesis that primitive forms of communication based on manual gesture preceded speech in the evolution of language. "Language readiness evolved as a multimodal manual, facial and vocal system with proto-sign, a manual-based protolanguage providing the scaffolding for proto-speech, a vocal-based protolanguage. There was a "neural critical mass" to trigger the emergence of language (Arbib, 2005, 2007; Kozma, Puljic, Balister, Bollobas, & Freeman, 2005) via the mirroring between neurons at the dendrite and axon levels. The neurodynamical result of this critical mass was the possibility to reach the threshold, in terms of dynamical systems number of degrees of freedom necessary for effective psychodynamics (Freeman, 1975; Orsucci, 1998). The *Mirror System Hypothesis* states that the matching of neural code for execution and observation of hand movements in the monkey is present in the common ancestor of monkeys and humans. *Imitation* plays a crucial role in human language acquisition and performance: brain mechanisms supporting imitation were crucial to the emergence of *Homo Sapiens*. Rizzolatti & Arbib (1998) hypothesize several stages of this evolution:

- Grasping
- A mirror system for grasping (i.e., a system that matches observation and execution)
- A simple imitation system for grasping
- A complex imitation system for grasping
- A manual-based communication system
- Speech, characterized as being the open-ended production and

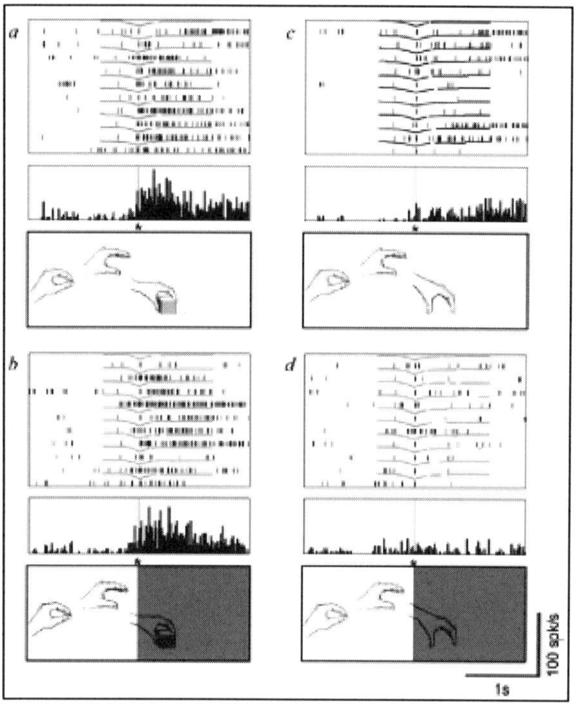

Fig. 33. Neurodynamics of mirror systems during an observed action (Rizzolatti et al., 1998).

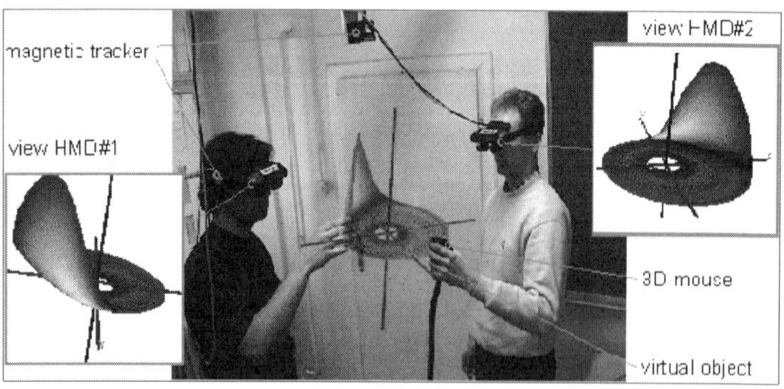

Fig. 34. Interacting with a physical-mathematical structure in augmented reality, the Roessler attractor (courtesy of Studierstübe).

perception of sequences of vocal gestures, without implying that these sequences constitute a language
- Verbal language

Following their findings and hypothesis, language evolved from a basic mechanism not originally related to communication: but from the mirror system with its capacity to generate and recognize a set of actions.

This new perspective generates some difficult questions posed by the interaction between new media and the mirror system. The different kinds of reality experiences produced by new media might activate, via direct perception, presentations or action-like brain effects, enduring plasticity effects. We are not referring just to the banal imitation induction one might experience after an immersive movie, but also to the longer lasting molding of the brain by the mirroring induced by all the most various contents provided by new media. It is a problem that older generations never encountered, and the spreading of diagnoses such as Attention Deficit Hyperactivity Disorder (ADHD) can be related to this.

The linguist Noam Chomsky (e.g., 1975; 1993) has argued that since children acquire language skills rapidly despite the "poverty of the stimulus", therefore the basic structures of language are encoded in the brain, forming a Universal Grammar encoded in the human genome. For example, it is claimed that the Universal Grammar encodes the knowledge that a sentence in a human language could be ordered as Subject-Verb-Object, Subject-Object-Verb, etc. By doing this the child simply needs to hear only a few sentences of his first language in order to "set the parameter" for the preferred order of that language. Others have argued against this theory, challenging instead that in fact the child has a rich set of language stimuli, and that there are now far more powerful models of learning than those that Chomsky took into account, allowing us to explain how a child might learn from its social interactions aspects of syntax which Chomsky would see as genetically pre-specified. Lieberman (1991) provides for a number of arguments which counter Chomsky's view.

Taken from the aforementioned arguments, we have observed that, for example, many youngsters today easily acquire the skills of "web surfing" and video-game playing despite a complete poverty of the stimulus, namely the inability of their parents to master these skills. We trust that no one would claim that the human genome contains a "web-surfing gene". Rather, we know the history of computers, and know that technology has advanced over the last 55 years to take us from an interface based on binary coding to a mouse-and-graphics interface so well adapted to human sensorimotor capabilities that a child can master it.

We reject Chomsky's view that many of the basic alternatives of grammatical structure of the world's current languages are already encoded in the human genome, so that the child's experience merely "sets parameters" to choose among pre-packaged alternative grammatical structures. The experimental evidence of this hypothesis, years after it was proposed, is still weak. The opposing view, which we support, holds that the brain of the first Homo sapiens was *language-ready* but it required many millennia of invention and techno-cultural evolution for human societies to form languages in the modern sense.

The structure of a language-ready brain had reached a critical neural mass action (Freeman, 1975) of connections and feedback redundancies capable to provide reflexivity and the emergence of consciousness. The mirror neurons finding is based on the massive increment of feedback and regulations embedded into the architecture of the human brain. In this sense, mirroring and reflexivity are embedded in the usual functioning of all neurons and structured in some more specialized ones. Chomsky and his followers instead, in some way, present a Platonist approach claiming that the so-called deep structures, symbols and genes, are primary and antecedent to bio-psycho-physical experiences. We prefer a more realistic complexity approach which recognizes different biological and non biological factors in language development (Orsucci, 2002a, 2002b; Tomasello, 2003).

In this framework, it is quite interesting to consider how Rizzolatti & Arbib (1998) propose that at Stage 5, the Manual-Based Communication System broke through the fixed repertoire of primate vocalizations to yield a combinatorial open repertoire. This was so that Stage 6, Speech, did not build upon the ancient primate vocalization system, but rather rested on the "invasion" of the vocal apparatus by collaterals from the communication system based on F5/Broca's area. In discussing the transition to Homo sapiens, they stress that our predecessors must have had a relatively flexible, open repertoire of vocalizations but this does not mean that they, or the first humans, had language skills. Rather, they hold that human language, as well as some dyadic forms of primate communication evolved from a basic mechanism that was not originally related to communication: the capacity to recognize actions.

Psychoanalytical studies highlight the important perspective of mirroring in emotional development. The reflexive function is central also in the definition of identity and relations. Sigmund Freud had focused on children at play after becoming famous as the *fort-da*, in which a mirror can be used by the child to represent the disappearance of the caregiver. Jacques Lacan (1937) proposed a specific stage in child development, called *le stade du miroir,* in which the child reaches recognition of her image in a mirror. This stage, linked to a crucial step in the integration of the Central Nervous System, is evident also in some primates and was considered crucial in the establishment of a self conscious identity. Eugenio Gaddini (1969, 1989) explored imitation as a primitive and primary form of identification.

Donald W. Winnicott (1958, 1971) extended this notion to reflexive responsiveness that a child can receive from the caregiver, the family and the extended social environment. Peter Fonagy (1997 & 2002) states that the reflective function is a developmental acquisition allowing the child to respond not only to other people's behavior, but to his own conception of their beliefs, feelings, hopes, pretence, plans, and so on. "Reflective function or mentalization enables children to *read* people's minds". Paulina Kernberg (2006) recalls how the mirror function of the mother is expanded to the idea of attunement between mother and child (Stern, 1985, 2004), resonating affectively, visually, vocally and by movement and touch.

Chapter 6

Evolutions

6.1 Mimesis

Judging from the anatomical and cultural remains left by hominids and early humans, the most important evolutionary steps were concentrated into a few transition periods where the process of change was greatly accelerated, and these major transitions introduced new capacities. Merlin Donald (1997), within the same research line, proposes to identify some evolutionary punctuation in the development of the human embodied mind. The first transition is mimetic skill and auto cueing. The rationale for the first transition is based on several premises:

- the first truly human cognitive breakthrough was a revolution in motor skill – mimetic skill – which enabled hominids to use the whole body as a representational device;
- this mimetic adaptation was a multimodal modeling system, and it had a self-triggered rehearsal loop;
- the sociocultural implications of mimetic skill are considerable, and could explain the documented achievements of Homo erectus;
- in modern humans, mimetic skill in its broadest definition is dissociable from language-based skills, and retains its own realm of cultural usefulness;
- the mimetic motor adaptation set the stage for the later evolution of language.

Mimesis can be just an emergent property of mass action in the nervous system, as the mirror function is a specialization of the arousal and feedback neural processes. The embodiment of mind processes becomes, in this way, a neurobiological necessity. As the whole body becomes a potential tool for expression, a variety of new possibilities enter the social arena: complex games, extended competition, pedagogy through directed imitation with a concomitant differentiation of social roles, a subtler and more complex array of facial and vocal expressions and public action-metaphor, such as intentional group displays (i.e. aggression, solidarity, joy, fear, and sorrow). The emergence of religious practice could be also considered, in its animistic beginnings, as an extension of mimetic functions to the living and non-living environment.

The second transition is the lexical invention. The rationale for the second transition is briefly as follows:

- since no linguistic environment had yet existed, a move towards language would have depended primarily on developing a capacity for lexical invention;
- phonological evolution was accelerated by the emergence of this general capacity for lexical invention, and included a whole complex of special neuronal and anatomical modifications for speech;
- the language system evolved as an extension of lexical skill, and gradually extended to the labeling of relationships between words, and also to the imposition of more and more complex meta-linguistic skills that govern the uses of words; and
- the natural collective product of language was a narrative thought, (essentially, storytelling) which evolved for specific social purposes, and serves essentially similar purposes in modern society;
- further advanced products are technical jargons and mathematical notations. These new representational acts, both speech and mimesis, are performed covertly as well as overtly.

Covert speech has been called inner speech or inner dialogue in order to stress how it is equivalent to the activation of the central aspects of articulation, without actual motor execution. The mental operation we call *imagination* can similarly be seen as a mimesis without motor execution of imagined acts and situations. The control of mimetic imagination (probably even of visual generative imagery, which is facilitated by imagined self-movement) presumably lies in a special form of kinematical imagery. Auto-recover is just as crucial for covert imaginative or linguistic thought as it is for the overt or acted-out equivalent. Thus, given a lexicon, the human mind was able to self-trigger recollection from its memory in two ways: by means of mimetic imagination, and by the use of word-symbols, either of which could be overt or covert.

The third transition is grammar and other meta-linguistic skills. According to the *competition* model proposed by Bates & MacWhinney (1987), the whole perisylvian region of the left hemisphere is diffusely dedicated to language, function words and grammatical rules being stored in the same tissue as other kinds and aspects of lexical entries (see also Bates, 1979). However, we readily admit that this issue, like many others in this field, has not yet been conclusively resolved; there is electrophysiological evidence that function words, those most relevant to grammar, might have a different cerebral representation from open-class words (Neville, 1992).

6.2 Time Sharing

In the classical sense, the word *synchronization* (literally from ancient Greek, "sharing time") means: "adjustment or entrainment of frequencies of periodic oscillators due to a weak interaction". Synchronization is a basic nonlinear phenomenon in physics, discovered in interactions between pendulums at the beginning of the modern age of science. More recently, Maturana & Varela (1980) had suggested that sync is *a form of structural coupling, a process which occurs when two structurally plastic systems repeatedly perturb one another's structure in a non-destructive way over a period of time*. This leads to the development of a structural 'fit' between systems. There is an intimate relationship between this process and the *emergence of 'appropriate' behavior from the interplay between interacting systems,* because the structure of a system determines its responses to perturbing environmental events. Humberto Maturana (2002) stressed this dynamical approach in semiotic terms within a co-evolutionary perspective: "Language is a manner of living together in a flow of coordination of consensual behaviors or actions that arise in a history of living in the collaboration of doing things together".

This dynamical systems approach leads to control and synchronization in chaotic or complex systems. As we have already seen,

Pecora & Carroll (1990) and Ott, Grebogi & Yorke (1990) discovered a new and reliable way to contemporary research on control and synchronization of complex systems.

We have been investigating sync during natural conversations, finding that it is a quite complex phenomenon occurring simultaneously with nonverbal, phonetic, syntactic and semantic levels (Orsucci, 1999, 2004, 2006). The statistical tool we consider most suitable for this kind of study is the Recurrence Quantification Analysis (Eckman et al., 1987; Webber & Zbilut, 1994; Zbilut, 2002; Marwan, 2003) Coordination between conversation partners occurs at multiple levels, including the choice of syntactic structure (Branigan et al., 2000). A number of outstanding questions concerning the origin of this coordination requires novel analytic techniques. Our research can be considered complimentary with a study by Shockley et al. (2003), in which interpersonal coordination during conversation was based on recurrence strategies to evaluate the shared activity between two postural time series in a reconstructed phase space.

In a study on speech and rhythmic behavior, Port et al. (1999) found that animals and humans perform many kinds of behavior where frequencies of gestures are related by small integer ratios (like 1:1, 2:1 or 3:1). Many properties like these are found in speech, as an embodied activity considered as an oscillator prone to possible synchronizations.

Our findings in the synchronization of conversation dynamics can be relevant for the general issue of structural coupling of psychobiological organizations. Implications are related with psycho-chrono-biology research and the clinical field.

Data on synchronization suggests that this dynamic behavior can be evident also in semiotic and cognitive dynamics, aside from the well established research on biological oscillators. For example, Dale & Spivey (2006) used this method to explore lexical and syntactic coordination between children and caregivers during conversation. Similar studies highlight synchronization of eye movements in conversations (Richardson & Dale, 2005).

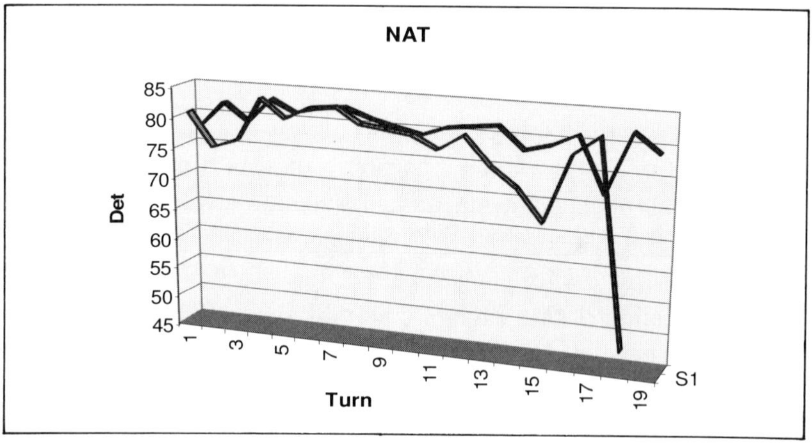

Fig. 35. Synchronization during a natural conversation (Orsucci et al., 2004).

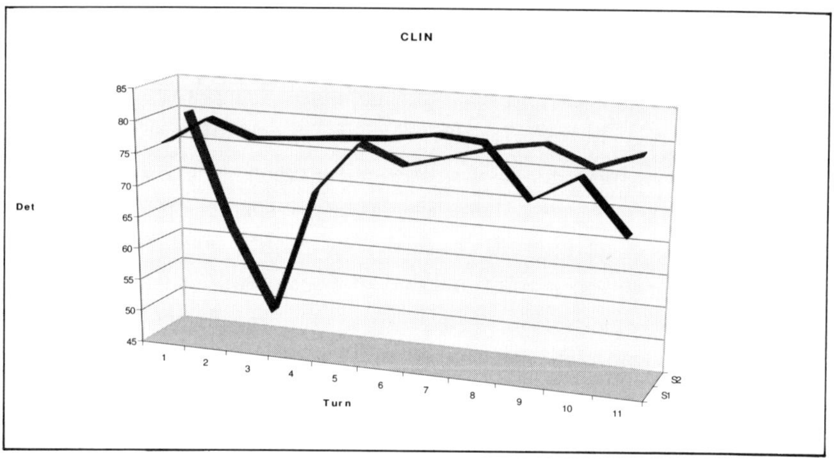

Fig. 36. Synchronization during a clinical conversation (Orsucci et al., 2004).

Synchronization is a crucial area of study that brings biophysics, neuroscience and information technologies together. Sharing time, in different timeframes, is critical for neurodynamics, consciousness and cooperation with humans and non-humans (machines included).

6.3 Enaction

Time, as we have seen, is a crucial factor in synchronization. Mirroring needs some coincidences, both in space and time, though their definition can be quite complex. Depending on the context, the framing of sync can change. Subjective, and neuro-cognitive, time in experience is quite different from time as measured by a clock. Time in experience presents itself not only as something that's linear but also as having a *complex texture* (evidence that we are not dealing with a *knife-edge* present), a texture that dominates our existence to an important degree (Varela & Petitot, 1999).

This overall approach to cognition is based on *situated embodied agents*. Varela (Varela et al., 1991; Thompson & Varela, 2001) has proposed the adjective *enactive* in order to designate this approach more precisely. It comprises two complementary aspects:

- Ongoing coupling of the cognitive agent, a permanent coping that is fundamentally mediated by sensory-motor activities;
- Autonomous activities of the agent whose identity is based on emerging, endogenous configurations (Kelso, 1995) of neuronal activity.

Enaction implies that sensory-motor coupling modulates, but does not determine, an ongoing endogenous activity that it configures into meaningful world items in an unceasing flow. From an enactive viewpoint, any mental act is characterized by the concurrent participation of several functionally distinct and topographically distributed regions of the brain and their sensory-motor embodiment. From the neuroscientist's point of view, the complex task of relating and integrating these different components is at the root of temporality. These various components require *a frame or window of simultaneity that corresponds to the duration of lived subjective present*. These kinds of present are not necessarily conscious. In fact, often they are not, though they might not be unconscious in the folk Freudian way (Orsucci et al., 2006). There are three possible scales of duration required to understand the temporal horizon just introduced (though other scales of extended present, considered in chrono-biology, might considered):

- basic or elementary events (the “1/10” scale);
- relaxation time for large-scale integration (the “1” scale);
- descriptive – narrative assessments (the “10” scale).

The first level is already evident in the so-called *fusion interval* of various sensory systems: the minimum distance needed for two stimuli to be perceived as non-simultaneous, a threshold that varies with each sensory modality. These thresholds can be grounded in the intrinsic cellular rhythms of neuronal discharges, and in the temporal summation capacities of synaptic integration. These events fall within a range of 10 ms (e.g., the rhythms of bursting inter-neurons) to 100 ms (e.g., the duration of an EPSP/IPSP sequence in a cortical pyramidal neuron). These values are the basis for the 1/10 scale. Behaviorally, these elementary events give rise to *micro-cognitive phenomena* variously studied as perceptual moments, central oscillations, iconic memory, excitability cycles, and subjective time quanta. For instance, under minimum stationary conditions, reaction time or oculo-motor behavior displays a multimodal distribution with a 30-40 ms distance between peaks. In average daylight, apparent motion or *psi-phenomenon* requires 100 ms.

This leads naturally to the second scale, a long-range integration. If component processes already have a short duration, about 30-100 ms, how can we then understand such experimental psychological and neurobiological results, at the level of a fully constituted, normal cognitive operation? A cell assembly (CA) is a distributed subset of neurons with strong reciprocal connections.

The upper diagram depicts the three main time frames considered here. A cognitive activity (such as head turning) takes place within a relatively incompressible duration, a *cognitive present*. The basis for this emergent behavior is the recruitment of widely distributed neuronal ensembles through increased frequently, coherence in the gamma (30-80 Hz) band. Thus, we might depict the corresponding neural correlates of a cognitive act as a synchronous neural hypergraph of brain regions undergoing bifurcations of phase transitions from a cognitive present content to another.

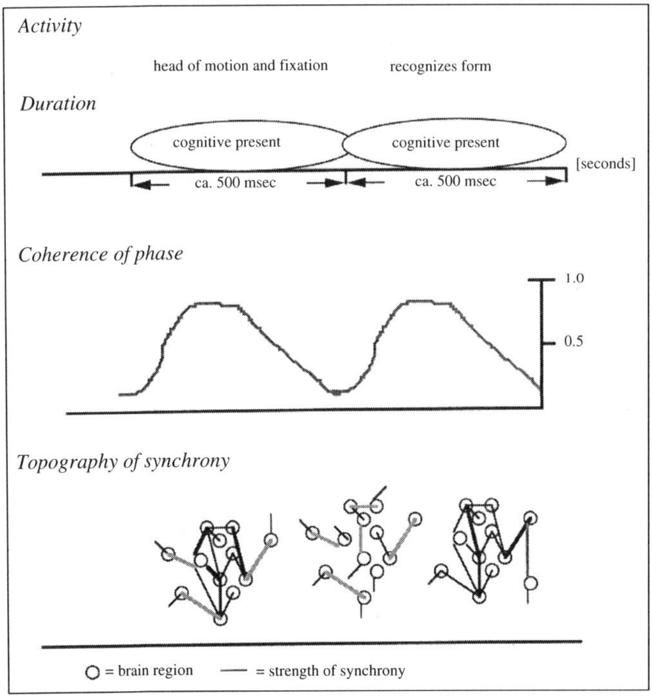

Fig. 37. Windows of time (Stewart & Varela, 1991).

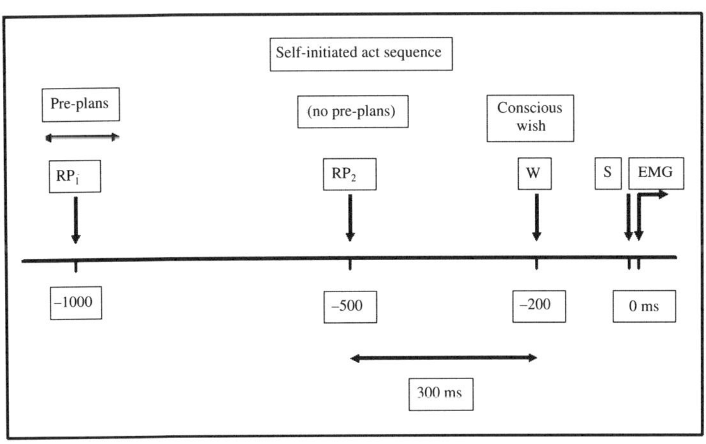

Fig. 38. Readiness potential (Libet et al., 1999).

This view has been supported by widespread findings of oscillations and synchronies in the gamma range (30-80 Hz) in neuronal groups during perceptual tasks. Thus, we have neuronal-level constitutive events that have a duration on the 1/10 scale, forming aggregates that manifest as incompressible but complete cognitive acts on the 1 scale. This completion time is dynamically dependent on a number of dispersed assemblies and not on a fixed integration period; in other words, it is the basis of the origin of duration without an external or internally ticking clock.

Nowness, in this perspective, is therefore pre-semantic in that it does not require a remembering in order to emerge. The evidence for this important conclusion comes, again, from many sources. For instance, subjects can estimate durations of up to 2-3 seconds quite precisely, but their performance decreases considerably for longer times; spontaneous speech in many languages is organized such that utterances last 2-3 seconds; short, intentional movements (such as self-initiated arm motions) are embedded within windows of this same duration.

This brings to the fore the third duration, the 10 scale, proper to descriptive-narrative assessments. In fact, it is quite evident that these endogenous, dynamic horizons can be, in turn, linked together to form a broader temporal horizon. This temporal scale is inseparable from our descriptive-narrative assessments and linked to our linguistic capacities. It constitutes the "narrative centre of gravity" in Dennett's metaphor (Dennett, 1991), the flow of time related to personal identity.

It is the continuity of the self that breaks down under intoxication or in pathologies such as schizophrenia or Korsakoff's syndrome. As Husserl (1980) points out, commenting on similar reasoning in Brentano: "We could not speak of a temporal succession of tones if … what is earlier would have vanished without a trace and only what is momentarily sensed would be given to our apprehension" To the appearance of the just-now one correlates two modes of understanding and examination (in other words, valid forms of donation in the phenomenological sense):

– remembrance or evocative memory
– mental imagery and fantasy.

The *Ur-impression* is the proper mode of nowness, or in other words, it is where the new appears; impression intends the new. Impression is always presentational, while memory or evocation is re-presentational.

These neurophysiologic events are correlated to micro-cognitive phenomena and behavioral elements variously studied as perceptual moments, central oscillations, iconic memory, excitability cycles, and subjective time quanta: the elementary particles of reflexions we can share with humans and media. Coupling and sharing between humans and machines are happening at this level, when meta-cognitive and mental skills are certainly unusual. It is the *a-conscious* level and modality, preliminary to any unconscious or preconscious modes. The kind of reflexivity implied in these processes concerns the embodied mind. It is a kind of cognitive capacity fully incorporated in bodily actions and re-actions. These kinds of processes involve a presentational intentionality, not a representational intellect. It is a form of direct cognition, not a self conscious meta-cognition.

The embodied mind emerges and grows (bottom-up) on the basic *reflexive function* as a direct parameter in biological processes. *Reflection* is a meta-cognitive function (top-down): "the overall reflective process can embed more conceptual and linguistic functions in the brain than the reflexive component alone" (Siegel, 2007). Some authors use the terms synonymously but, we prefer to use a different terminology to stress a conceptual and factual difference. Reflexivity will be direct and non-conceptual: it implies an immediate capacity of awareness without effort, or intellectualization. In reflexivity the interface is just like your own skin. It is useful to remind that the embryological origin of skin, brain and mind are the same. The *ectoderm*, our primary interface, is the outermost of the three primary germ layers of an embryo and the source of the epidermis, the nervous system, the eyes and ears: i.e. interfaces.

6.4 Readiness Potential

Reflexions happen at a very pre-cognitive stage, before any higher order metacognition might be established. We have been exploring some important implications of mirror neuron research. The findings by

Benjamin Libet (1993, 1999) on the so-called readiness potential extend our perspectives on this matter. Kornhuber & Deecke (1965) had found that all actions are preceded by a slow potential easily detected in an EEG. They gave this potential a German name, *Bereitschaft-potential* but nowadays it is more frequently called in English: *Readiness Potential or RP*. A question was quite immediate: as the RP was happening at about 550 milliseconds before action, in which timing (and maybe causal) sequence was it placed with representations and decisions concerning that same action? It was found that every conscious representation and decision of acting were placed at just 200 msec. before action, so they were following the RP at about 300 msec. Benjamin Libet, an American neuro-physiologist has been expanding research in this area both in the sensory and the motor fields, including some possible psychological and philosophical implications (Libet, 1993, 1999). His vantage point might be summarized in this way: we don't have free-will but we do have free-denial. We have the possibility to facilitate or stop an action after it has been started, in an *aconscious* way. Recent advancements in the complex neurodynamics of time could provide seminal contributions in advancing our understanding of ethical issues on personal responsibility of actions (De Risio, & Orsucci, 2004; Gazzaniga, 2005).

It is not surprising then that some recent research is showing evidence on how new media have a direct neurocognitive impact, including probable long term evolutionary results (Chan & Rabinowitz, 2006). New media are exploiting our physiological capacity of direct sensation and reaction, through aconscious interactions. New media constitute a techno-cultural niche, an enriched or enhanced environment, based on forms of direct knowledge. A knowledge not mediated by intellectual representations.

Every generation has raised concerns regarding the negative impact of media on social skills and personal relationships. The Internet and other new media types are reported to have important social and mental health effects on everyone, especially on adolescents probably because they are heavy users and their brains and psychology are still very moldable.

Video-game playing, for example, enhances the capacity of visual attention and its spatial distribution. Video-game training enhances task-switching abilities as well as decreasing the attention blink. Thus, both the visual and a-modal bottlenecks identified during temporal processing of visual information are reduced in video game players. Clearly, these individuals have an increased ability to process information over time; however, whether this is due to faster target processing, such as faster selection and stabilization of information in memory, or to an increased ability to maintain several attention windows in parallel, cannot be determined from our current data.

By forcing players to simultaneously juggle a number of varied tasks (detect new enemies, track existing enemies and avoid getting hurt, among others), action-video-game playing pushes the limits of three rather different aspects of visual attention. It leads to detectable effects on new tasks and at untrained locations after only 10 days of training. Therefore, although video-game playing may seem to be rather mindless, it is capable of radically altering visual attention processing. There are several ways by which video-game training could lead to such enhancements. Changes in known attention bottlenecks is certainly a possibility; however, speeded perceptual processes and/or better management of several tasks at the central executive level are also likely to contribute. It will be for future studies of the effect of video-game practice to determine the relative contribution of these different factors to skill learning (Green & Bavelier, 2003).

For example, has been reported a statistical association between television viewing and obesity, attention disorders, school performance, and violence (Mathiak & Weber, 2006). A significant relationship between Internet use and attention deficit hyperactivity disorder (ADHD) has also been shown in elementary school children (Yoo et al., 2004). The relationship between video games and Attention Deficit Hyperactivity Disorder is unknown. The incidence of ADHD continues to rise and it is a significant challenge on medical, financial, and educational resources. ADHD is a complex disorder that often requires input from the affected child or adolescent, teachers, parents, and physicians in order to be diagnosed correctly and treated successfully. Adolescents who play more than one hr of console or Internet video games a day may have more or more intense symptoms of ADHD or inattention than those who do not (Straker et al., 2006).

6.5 Enriched Environments

New media, new reflexions, are already part of the techno-cultural niche of our age: this is part of a new evolutionary step we can better understand considering studies on learning and enriched environments. At Harvard, David Hubel and Thorsten Wiesel studied cats raised blind in one eye, and by 1962 they had demonstrated that such deprivation caused profound structural changes in the cats' visual cortex. Hubel and Wiesel's work made it clear that severe deprivation during critical developmental periods could have catastrophic effects on a growing brain, but the question of whether the opposite was true remained suspended for a while.

By 1964, the Berkeley team led by Mark Rosenzweig completed a series of experiments that began to answer those questions. They found that rats raised in an "enriched" environment, with toys and interesting social activities, were not only smarter than rats raised in impoverished environments, but that the improvement in performance correlated with an increase in the weight of the rats' cerebral cortex. More recently Lamberto Maffei and his research team have clarified how enriched environments can influence sight development or recovery (Sale, 2004; Cancedda, 2004). The idea that the brain, like a muscle, might respond to "cerebral exercise" with physical growth was surprising to many, and gave strength to an increasingly powerful theory suggesting that all aspects of the mind — from memory, to dreams, to emotions — have physical correlates.

The classical statement by William James (now in 1967) has found an experimental validation: "Experience is remolding us at every moment: Whilst we think, our brain changes". Studies on enriched environments are still growing, but they have already established evidence that the brain modifies its structure (not necessarily its size) depending from the kind of niche that Rosenzweig called Environmental Complexity and Training. These studies are now extended to human learning environments (Carbonara, 2005; Orsucci & Sala, 2005).

Our new insights in neuro-cognition and the multiple reflexions implied in our sensory-perceptive processes are leading to new interfaces and new media. The isomorphism between bio-cognitive structures and the ICST niche we inhabit is progressively blurring boundaries between *res cogitans* and *res extensa*. Reflexing interfaces are extensions of human embodiment just as the bone tool tossed skyward by a savannah-dwelling ape. Time flows, always different yet similar. As Francisco Varela distilled aphoristically: "Readiness-for-action is a micro-identity and its corresponding level a micro-world: we embody streams of recurrent micro-world transitions" (1991).

Chapter 7

Attractions

7.1 Interpersonal Forces

The presence of strange and specific patterns of human attractions which might be due to what we came to call Mind Force came to our attention for empirical and theoretical reasons.

From the empirical point of view, beyond the studies already cited, there are several papers in which we published (Orsucci, 1996b, 2006) clinical observations concerning the process of merger, sharing and emergence leading to mind changes.

These concurring theoretical, empirical and clinical interests were directing our research to the dynamic role that might have in a standard relationship, and in psychotherapy, which is a specialized form of relationship, to the immense bulk of communication stream involved, most of it made of body information. A stream exceeding by far the explicit verbal content we are used to study in standard psychotherapeutic and psychiatric training. Sigmund Freud was acquainted with this problem when he used to stress that every verbal act has an over-determined meaning, made of intermingled layers of experience, thinking and culture; some private, some shared. Unfortunately he and many of his followers decided to simplify and give a deterministic version of his technical instructions. At the same time, as we will see, other psychoanalysts were trying to describe, in some sort of philosophical and conceptual terms the relational force fields and its evolutions. We might now try to follow some embryonic intuitions on Mind Force within the history of the psychoanalytical movement.

Freud has had a long correspondence with a friend, Wilhelm Fliess, which is considered to be an incubation of the construction of psychoanalysis and many works he published later in his life (Freud, 1985). Included in this correspondence there is a rather long manuscript now called "G Draft, Melancholy". In the manuscript we can find a scheme which is strangely similar to a phase space, in physical terms, and it represents the trajectory of a vector describing an interpersonal, sexual, relationship. The elements of the model are allocated in a space divided by the intersection of two boundaries (the Ego border, and the psychosomatic border). It is amazing how this dynamical and complex model has been neglected in the history of psychoanalysis and psychosomatics. This model is clearly dynamic, while other famous Freudian models are more deterministic and static. It is probable that this model has influenced by the psychophysics school, as Freud had been trained in the school of Weber and Fechner (Orsucci et al., 1999).

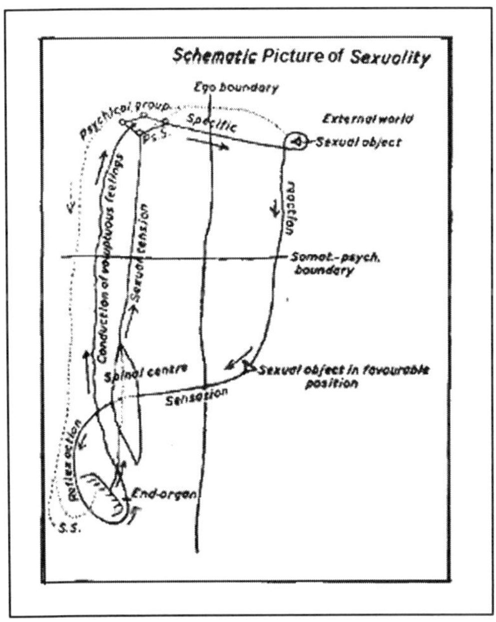

Fig. 39. The G-Draft scheme (Freud, 1985).

Melanie Klein, one of the main psychoanalysts of the 40s, had a strong intuition of the way MF might forge interpersonal dynamics. Her main contribution to the psychoanalytic theories should be considered the concept of *projective identification*, which she considered a psychological way a person might use to expel unpleasant fantasies and emotions and push them within a partner. This concept was also a mix of two classical Freudian concepts: 1) identification, a way to find a similar trait in another and by this way to assimilate the reciprocal identities and 2) projection, a way to push in fantasy and language, parts of own identity to the outside reality and partners. She considered this way (which in modern terms we might consider as a distortion of the mirror function) as an aggressive way to express "an omnipotent fantasy to control the other from the inside" (Cit). Klein's proposals never received a scientific testing though she was the leader of one of the key post-Freudian schools. She was also criticized because she reached the maybe paradoxical conclusion that the projective identification was just a pathological way to stay in a relationship and the correct clinical strategy was to "tame" and reduce it.

Wilfred Bion, one of her most important followers, reached a more balanced position by recognizing that the projective identification can be also *realistic*: a way to test emotional reactions in partners and a primitive form of empathy. He gave us inspiring clinical descriptions, introspections and speculations (1992) in which he presented his acute perception of the immense heterogeneity of components in mind/body processes.

Donald W. Winnicott, another contemporary master in psychoanalysis, was the discoverer of transitional objects, later popularized in many different ways. He recognized in transitional objects and phenomena the satellites of the main love relations and a way for a personal development from attachment to social involvement.

7.2 Bipersonal Fields

More recently the psychoanalytic context has provided a very interesting contribution, proposed by Madeleine and Willy Baranger. It can be considered within the Mind Force framework: the *bi-personal field theory*. These authors evolved the Kleinian construct by including contributions from Gestalt psychology and Maurice Merleau-Ponty (Borges, 1964) phenomenological psychology. The result has been quite new and original. They describe their approach stressing that analyst and patients always share the same dynamical process: they are immersed within the same field, which should be considered as a third element, a medium, in the psychoanalytic process. They propose that within the *bipersonal field* operates what they call an unconscious bi-personal fantasy, produced by the crossover of reciprocal projective identifications. The bi-personal fantasy, in their opinion, is the core of a psychoanalytical process and it usually develops after activation of the dynamical field. This process would mobilize blocked projections and introjections, a blocking which could be considered a cause of mental suffering.

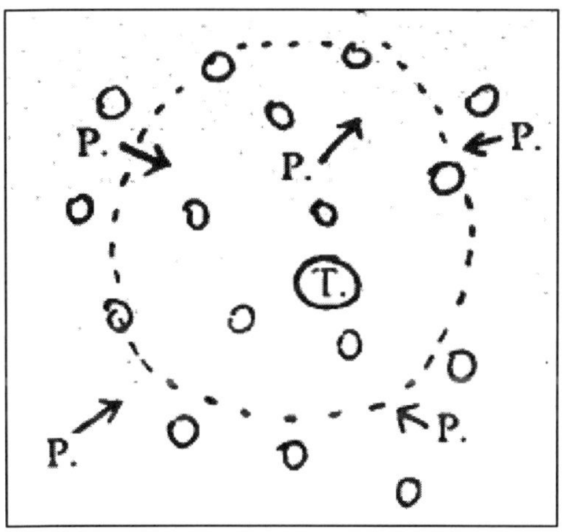

Fig. 40. Bipersonal field in a therapeutic situation (Winnicott, 1958).

The bipersonal field theory brings a remarkably non deterministic approach in considering subject and object as embedded in a dynamical field and producing a third, non-Cartesian element, the bipersonal fantasy.

The result is an important step towards recognition of MF force lines whose derivatives might be very heterogeneous: from verbal to non verbal, extra verbal or ultra verbal (Corrao, 1986, 1992; Neri, 1993, 1998).

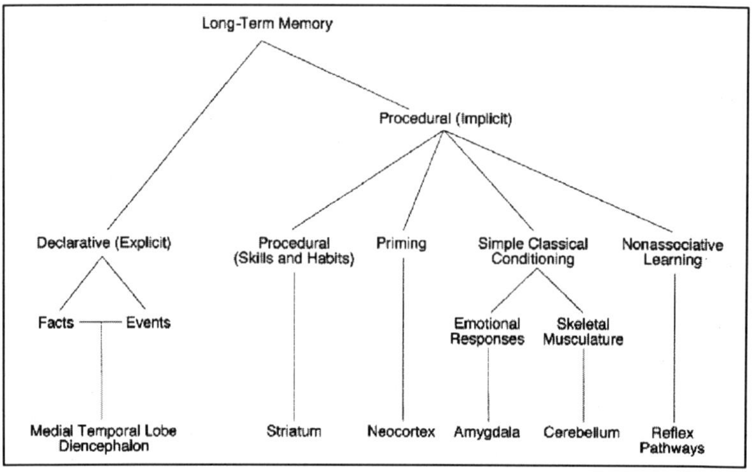

Fig. 41. Different kinds of memory (Kandel, 1999).

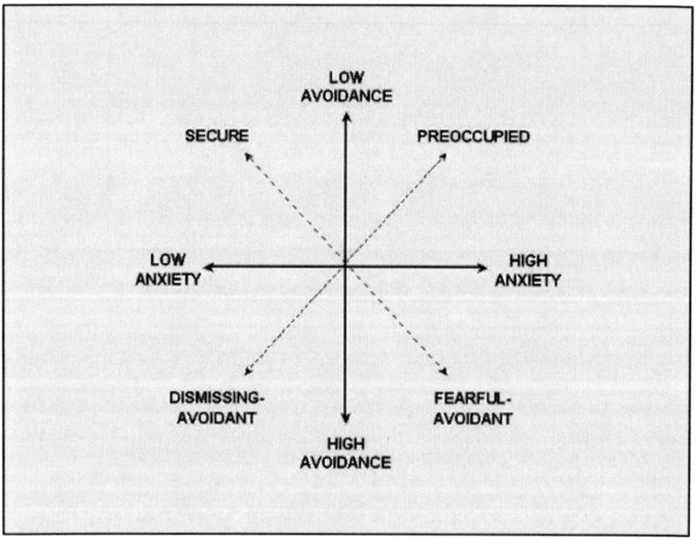

Fig. 42. Motivational system and personalities in attachment theory.

A further crucial step along this line of a potentially radical reconsideration of psychotherapies is proposed by Daniel Stern when he focuses on the *present moment* and its role for the subject and the therapeutic relationship. He states that 'the sharing creates a new intersubjective field between the participants that alters their relationship and permits them to take different directions together. The moment enters a special form of consciousness and is encoded in memory. And, importantly, it rewrites the past. Changes in psychotherapy (or any relationship) occur by way of these nonlinear leaps in the way-of-being-with-another.' (Stern, 2004: 22).

It is evident in Stern's contribution an interesting synthesis from recent advancements in neuroscience (procedural memory and mindfulness), and resonances both from nonlinear dynamics and phenomenology. He explicitly cites Husserl, William James, Merleau-Ponty, Kandel and Varela. The result is what he calls *intersubjective matrix*, 'a mutual interpenetration of minds that permits us to say, "I know that you know that I know". It is a big step from a one person psychology to a bi-personal or even a multi-personal psychology. Stern is finally stressing the importance of *implicit knowledge* in intersubjective dynamics. 'Implicit knowledge is non-symbolic, nonverbal, procedural, and unconscious in the sense of not being reflectively conscious. Explicit knowledge is symbolic, verbalizable, capable of being narrated, and reflectively conscious' (2004: 113). The importance of implicit knowing poses a major challenge for traditional psychoanalysis and the whole area of psychotherapies mostly based on verbal language and narration. Implicit knowledge is a-conscious, non verbalizable and embodied.

7.3 Intersubjective Matrix

Myron Hofer has provided basic research on the neurobiology of the intersubjective matrix: a biopsychology of coupling in animal models of similar deep relations in humans (infant-mother, lovers etc.) (Hofer, 1994a).

Following Bowlby's observations (Bowlby, 1978, 1982) on attachment behavior, psychosomaticists studied the mother-infant bond as a strong coupling organization regulated mainly by its "inner" emotional signals. The strength of this coupling make it so protected and differentiated from influences coming from the environment, to make it an ideal sample for the study of psychobiological couplings. In a series of studies on rodents and primates, Myron Hofer and other developmental biologists demonstrated many hidden regulatory mechanisms (Hofer, 1984, 1994a, 1994b). These hidden multiple and pre- and intra-emotional factors act on different sensorial channels: nutritional, olfactory, tactile, thermal, visual and vestibular.

For example, the importance of bodily contact and tactile stimulation was demonstrated also by findings that a decrease in the levels of growth hormone, GH, in separated rat pups can be prevented by stroking their skins with a brush. Also, a 30% reduction of heart rate following separation could be prevented by providing the pups with a feeding. The body temperature of infant rats, which is determined largely by the body temperature of the mother, has been shown to regulate levels of brain peptides, nucleic acids and neuro-amines. All of them are reduced, if the young rats are prematurely separated from their mother.

Olfactory stimuli are also involved in the regulation of crucial aspects: for example infant rats are unable to locate the nipple in absence of a pheromone secreted from the mother's areolar glands, whose secretion is stimulated by suckling (interesting learning recurrence).

There is evidence of a multiplicity of regulatory mechanisms, most of them hidden for a passive third-observer, outside of the "nursing-couple". These regulatory mechanisms can be discovered either in experimental contexts or in pathologies implying an alteration of their normal functioning. In humans these kinds of regulatory mechanisms are operated on many different levels besides overt emotional expression and verbal language. They are supposed to continue in adult life, when their functioning becomes intermingled with other emotional, cognitive and social factors. They play a basic role in growth and health during all lifespan.

More recently, Jan Winberg (2005) reviewed 30 years of scientific research showing how early interactions between mother and newborn infant influence the physiology and behavior of both. Close body contact helps regulate the newborn's temperature, energy conservation, acid–base balance, adjustment of respiration, crying, and nursing behaviors. Similarly, babies may regulate mother's attention to his/her needs, breastfeeding, and the efficiency of her metabolism through vagus activation and a surge of gastrointestinal tract hormone release resulting in better exploitation of ingested calories.

7.4 Mediators and Genes

Researchers from Emory University, Atlanta, with colleagues in Boston and Tallahassee, Fla., found that boosting the expression of one gene switched meadow voles, a type of rodent, from a promiscuous to a monogamous lifestyle (Winslow et al., 1993; Insel et al., 1997). Larry J. Young says it's possible that the gene which encodes the *vasopressin V1a* receptor lies in a pathway homologous to one underlying human love.

Bartels and Semir Zeki of University College London ran functional magnetic resonance imaging (fMRI) scans of couples who professed to be deeply in love, as they viewed photographs of their beloveds. The researchers found significant activations in brain networks rich in receptors for the hormones *vasopressin* and *oxytocin*. These are the hormones identified as crucial for pair bonding in a series of animal studies, including Young's work with voles, suggesting the pathways are homologous (Bartels & Zeki, 2000).

An observation that may lead to greater clarity is that brain regions rich in oxytocin and vasopressin receptors overlap strongly with those rich in dopamine, the neurotransmitter classically associated with the brain's reward system. Young and colleagues propose that long-term partner preference occurs when the vasopressin circuits, which are also known to mediate individual recognition, somehow connect with the dopamine pathway, causing an animal to associate a specific individual with a sense of reward. In key dopamine-rich brain regions, vasopressin V1a is expressed more highly in monogamous than in promiscuous mammals. It seems that the mechanism of attachment preference uses the dopamine pathway to make attachment a rewarding experience.

Attachment is one of three neural systems Fisher et al. (2002) distinguish in connection with reproduction. The other two are *romantic love*, characterized by increased energy and focused attention on a preferred partner, and *lust*. Each can operate independently, and is associated with a different neurochemistry:

- *attachment* with vasopressin and oxytocin, and endorphins (Panksepp, 1998, 2004);
- *love* with dopamine and norepinephrine;
- *lust* with testosterone.

In prairie voles they have been studying attachment because the research focused on the creatures' lifelong bonds. Like Bartels, Fisher has conducted neuroimaging studies of lovers as they view photographs of their beloveds. Her results highlighted brain areas associated with dopamine and norepinephrine, in patterns similar, though not identical, to those Bartels found. Both say the differences might have arisen because Fisher's studies required that participants had just fallen in love, whereas Bartels' studies didn't impose this requirement.

Thus Bartels could have picked up more signs of what Fisher calls attachment.

7.5 Bonding

In interesting continuity, Walter J. Freeman (2001) focuses on the complex biological and cognitive techniques for inducing change in attachment patterns, a process which he calls as *dissolution*. Individuals separate themselves or are isolated from their normal social surroundings and support systems. They engage in or are subjected to severe physical exercise as in dancing, sports, and military drills, lack of sleep, and chemical stresses of their brains through the induction of powerful emotional states of love, hate, fear or anger. At some threshold the customary structure of the individual begin to crumble, and a collapse may occur that was described by Ivan Pavlov as *transmarginal inhibition*, the stage of physiological arousal beyond which further excitation leads to paradoxical depression. The experience may range from ecstatic through degrees of elation or discomfort to the stark terror of psychic free fall (Sargant, 1957). There is regression to successively earlier levels of *assimilation* as the structure of intentionality dissolves, particularly with resurfacing of old patterns of relations to parental care. There is a loss of normal constraints on behavior, and, in extreme instances, of language, locomotion, posture and even consciousness as the individual collapses. Recovery from *collapse* is followed by a state of extreme suggestibility, in which the skills of language and the competencies of daily living are regained, yet new values and habits can be established. This is done in a social setting of care by attendants who induce by example and exhortation the cooperative behaviors with the new companions and the social organization they embody and provide. This is a two-way process, because the caregivers get strong feelings of satisfaction from their supportive actions, and the recipients have strong sensitivity to peer pressures experienced as feelings of need for approval. The process is frequently referred to as being *reborn* (Verger, 1954). In the absence of support there is re-establishment of the *status quo ante*, meaning that the opportunity for change can be lost, attesting to the high degree of dynamic stability that characterizes intentional structures in normal circumstances.

The most likely candidate for a leading role in dissolution is a chemical neuromodulator named *oxytocin* (Pedersen et al., 1992). This neuropeptide has been known for many years as the agent in the female body that induces labor in parturition and subsequently lactation. More recently oxytocin has been found to be released by the brain into itself during sexual intercourse, particularly during orgasm in both men and women, and to be implicated in pair bonding not only of the parents to the child but also of parents to each other. The neurochemical actions of oxytocin in the brain are widespread, extremely complex, and difficult to study, so that much remains to be explored, but present knowledge shows that this neuropeptide is capable of inducing the meltdown of past learning that enables new learning. A simple example is the release of oxytocin flooding the brain of the multiparous ewe during delivery of her second and later litters, following which the dam refuses to nurse her earlier litters, having expunged the olfactory imprint required for maternal recognition of them as her offspring (Kendrick et al., 1992).

This primitive but well documented instance of dissolution serves also to explain its biological utility. Oxytocin is not likely to act alone, rather in concert with other neuropeptides all known to mediate states of emotion and levels of affect and disposition (Panksepp, 1998; Pert, 1997).

7.6 Affiliation

Freeman postulated that *affiliation* (Carter et al., 1997) is realized through new learning by cooperative behaviors driven by brains that have been prepared by the neurochemical changes precipitating dissolution, a state transition that leads to regression and clears the way to formation of new brain circuits. Cooperative action is the bedrock of social bonding: inadequacy of knowledge is compensated by the development of blind trust, which transcends language. The social technology of bonding is well known, having been explored by anthropologists in studies of tribal rites of passage, ordeals, and ceremonies (Verger, 1954), often accompanied by use of music, drumming, dance, and other forms of predictable repetitive actions, and by symbols such as flags, icons, totems, and, in modern times, corporate logos, military insignias and even tattoos. Dissolution may be occurring in critical phases of everyone's life and perhaps in minimal degree every night during sleep and dreaming.

In their recent review (2005) on a neurobehavioral model of affiliative bonding, Richard Depue and Jeannine Morrone-Strupinsky, propose a differentiation of three different systems regulating affiliative behaviors, social bonding and romantic love:

- a *dopamine* system, regulating appetitive processes and incentive rewards;
- an *opiate* system, regulating consummatory processes and reward;
- a system involving gonadal *steroids*, *oxytocin* and *vasopressin*.

You might notice that in this case the tripartite is quite similar to the other one, mentioned in the previous paragraph. The only difference is that oxytocin and vasopressin are included in a different system. It is a sign that, though their importance is widely recognized, their functioning and interactions with other neuromediators are very complex and still under scrutiny.

We examined some psychoanalytical, cognitive and neurobehavioral perspectives on the way people are connected. The integration of these different perspectives can be reached by considering Mind Force as the global framework. In order to complete the picture we are going to consider some models on love connections and, in the next chapter, social connections.

7.7 Love Dynamics

Romantic relationships are somehow the simplest case since they involve only two individuals. One of the most reliable approaches to the problem is rooted in attachment theory (Bowlby, 1969, 1973, 1980) which explains why infants become emotionally attached to their primary caregivers and why they often experience emotional distress when separated from them. Empirical research has focused on different attachment styles in children (Ainsworth, 1969, 1985) and adult individuals (Fig. 42). Several hypotheses have been generated about the nature and emotional quality of romantic relationships developed by people who exhibit different attachment styles.

Attachment styles can be regarded, in biophysical terms, as the different approaches of each human system to coupling and subsequent phase transitions in synchronization and co-evolution.

In conclusion, one can reasonably argue that the main features of love-stories should be largely dictated by the attachment styles of the individuals involved. Love-stories are dynamic processes that start from zero (two persons are completely indifferent to each other when they first meet), develop (more or less quickly) and end up into some sort of regime. Real-life observations tell us that most of the times transients develop very regularly and asymptotic regimes are stationary and associated to positive romantic relationships. But there are also love-stories initially characterized by stormy patterns of the feelings as well as by cyclic regimes, like that identified by Jones (1995) in Petrarch's *Canzoniere*, his celebrated book of love poems. This reminds very much the behavior of dynamical systems tending toward equilibriums or limit cycles. We have discussed in detail these dynamical landscapes in human attractions in a previous book, *Changing Mind* (Orsucci, 2002).

Observations point out the existence of multiple attractors. For example, it is known that steady and high quality romantic relationships can turn into a state of permanent antagonism after a disturbance, for example after a temporary infatuation of one of the two partners for another person. Bifurcations can be naturally invoked if one looks at the effects of age and other social variables (power, money etc.), capable of transforming tempestuous relationships into steady ones or vice-versa.

Strogatz (1988) first suggested in a 1-page contribution and later in his book (1994) analyzed love affairs via differential equations. This unusual approach was picked up by an Italian–Austrian group (Gragnani et al., 1997; Rinaldi, 1998a, 1998b) for a serious try to model such relationships by simple mathematical models composed of ordinary differential equations but under a bit more realistic circumstances. Recently another attempt (Sprott, 2004, 2005) was made to explain also the dynamics of love triangles. Related discrete dynamical models have recently been proposed by Gottman et al. (2002).

The dynamics of romantic relationships of two individuals are significantly more complex when time fluctuations of source terms and system parameters are taken into consideration. It is clear that time-independent personalities and appeals are an idealization. In reality, there are short- and long-termed fluctuations of personal feelings due to, for instance, biological cycles and varying stress life events. While it is straightforward to introduce such emotional patterns into the mathematical model, it is difficult to measure parameter values quantitatively. For couples with all kinds of exotic feeling, a large number of phenomena may be expected. However, the variability is expected to be more limited for couples of cautious and/or steady individuals.

One interesting effect is the cyclic change of love about a stationary center of steady love for both centers with alternating love and hate of different time spans might appear during different time periods, but it seems very seldom for cautious lovers with strong damping.

Closing the discussion, it may be noted that the examination of love triangles with time-varying fluctuations is another example of particular interest. Assume a couple governed by linear state equations, or a so-called robust couple governed by equations with nonlinear return functions. For a fragile couple, there is a second stable fixed point of stationary hate, and the cyclic behavior may occur about either of the two centers. A more complex irregular motion pattern surrounding both centers with alternating love and hate of different time spans might appear during different time periods, but it seems very seldom for cautious lovers with strong damping. Wauer et al. (2007) recently showed how linear and nonlinear robust couples of cautious individuals with time-fluctuating feelings tend to a cyclic emotional behavior about a fixed point of love while for nonlinear fragile couples, more complex patterns of emotional feelings are possible.

The artistic intuitions of Goethe's *Elective Affinities* can finally find a formal modeling and the dynamical landscapes of Mind Force in love can be further explored.

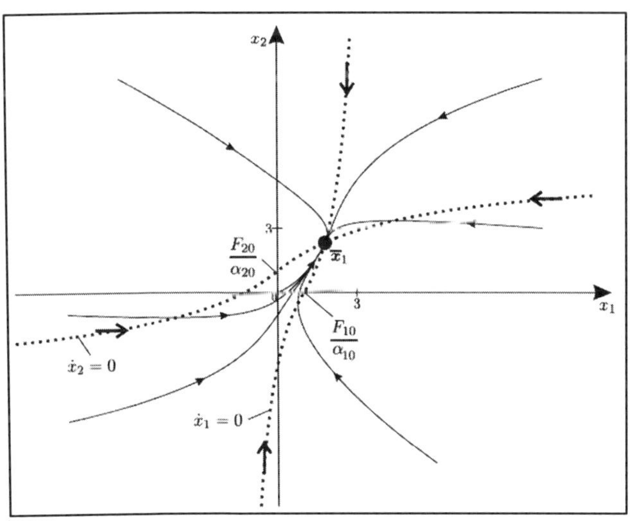

Fig. 43. Evolution of the feelings in a robust couple for different initial conditions (Wauer et al., 2007).

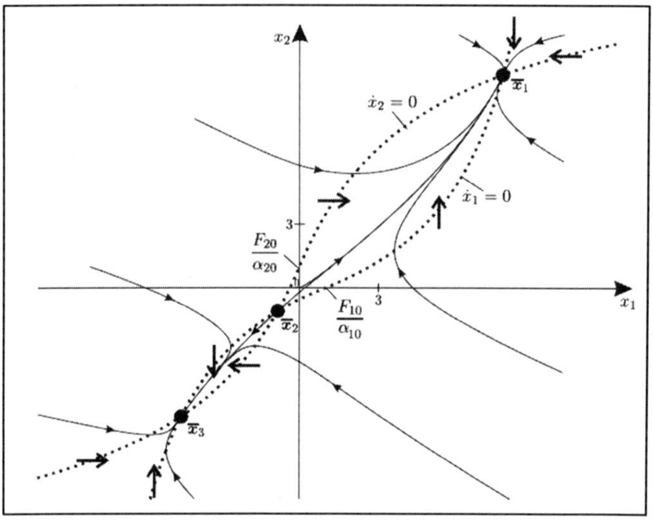

Fig. 44. Evolution of the feelings in a fragile couple for different initial conditions (Wauer et al., 2007).

Chapter 8

Societies

8.1 New Wave

The modern theory of networks, which originated with the discovery of small-world networks and scale-free networks at the close of the last millennium, is the most recently developed approach to complex systems.

The study of complex networks has attracted a large amount of attention in the last few years, and has resulted in applications in such various fields as the study of metabolic systems, airport networks and the brain. The modern theory of networks has its roots in mathematics as well as in sociology. In 1736 the famous mathematician Leonard Euler (1707–1783) solved the problem of the *bridges of Konigsberg*. This problem involved the question whether it is possible to make a walk crossing exactly one time each of the seven bridges connecting the two islands in the river Pregel and its shores. Euler proved that this is not possible by representing the problem as an abstract network: a *graph*. This is often considered the first proof in graph theory. Since then graph theory has become an important field within mathematics, and the tool to handle network properties theoretically.

An important step forward occurred when random graphs were discovered. In random graphs connections between the network nodes are present with a probability p. Many important theorems have been proven for random graphs. In particular it has been shown that properties of the graphs often undergo a sudden phase transition as a function of increasing p.

This phenomenon was probably first observed by the Hungarian writer Frigyes Karinthy in a short story. In this story he speculates that in the modern world the *distance* between any two persons is unlikely to be

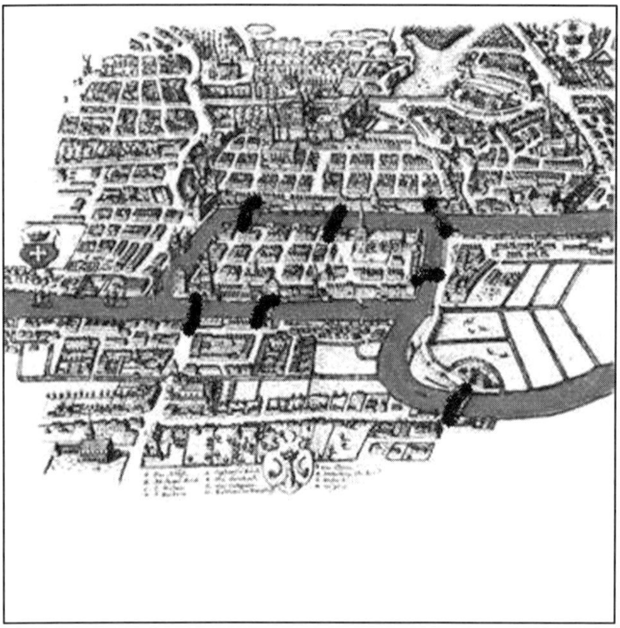

Fig. 45. Seven bridges of Konigsberg.

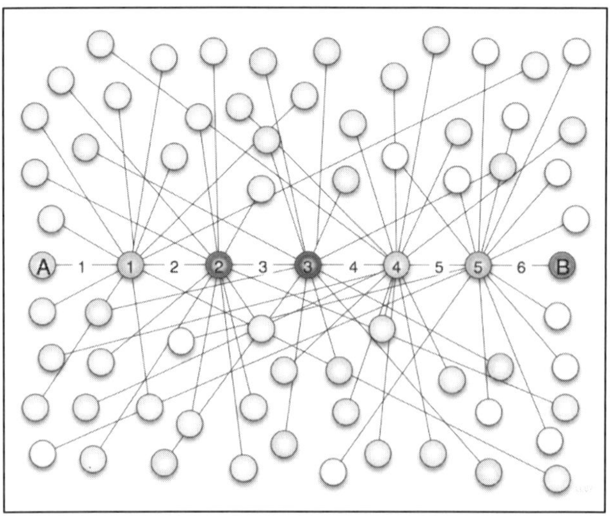

Fig. 46. Six degrees of separation.

more than five persons. As it turned out, this was a remarkable foresight of an important fact about certain classes of networks. The first person to study this phenomenon more scientifically was Stanley Milgram (1933–1984). While he was in the late 1960s at Harvard University (Milgram, 1967), he performed one of the first quantitative studies of the structure of social networks.

He took a number of letters addressed to a stockbroker acquaintance of his in Boston, Massachusetts, and distributed them to a random selection of people in Nebraska. His instructions were that the letters were to be sent to the stockbroker by passing them from person to person, and that, in addition, they could be passed only to someone whom the passer knew on a first-name basis. Since it was not likely that the initial recipients of the letters were on a first-name basis with a Boston stockbroker, their best strategy was to pass their letter to someone whom they felt was nearer to the stockbroker in some social sense: perhaps someone they knew in the financial industry, or a friend in Massachusetts.

A reasonable number of Milgram's letters did eventually reach their destination, and Milgram found that it had only taken an average of six steps for a letter to get from Nebraska to Boston. He concluded, with a somewhat cavalier disregard for experimental niceties, that six was therefore the average number of acquaintances separating the pairs of people involved, and conjectured that a similar separation might characterize the relationship of any two people in the entire world. This situation has been labeled *six degrees of separation* (Guarc, 1990), a phrase which has since passed into popular folklore.

Given the form of Milgram's experiment, one could be forgiven for supposing that the figure six is probably not a very accurate one. The experiment certainly contained many possible sources of error. However, the general result that two randomly chosen human beings can be connected by only a short chain of intermediate acquaintances has been subsequently verified, and is now widely accepted (Korte & Milgram, 1970).

8.2 Small-Worlds

In the jargon of the field this result is referred to as the small-world effect. The small-world effect applies to networks other than networks of friends. Brett Tjaden's parlor game *The Six Degrees of Kevin Bacon* connects any pair of film actors via a chain of at most eight co-stars (Tjaden & Wasson, 1997). With tongue very firmly in cheek, the New York Times played a similar game with the names of those who had tangled with Monica Lewinsky (Kirby & Sabre, 1998).

But, for a long time no satisfactory explanation was available though networks are ubiquitous. The brain is a network of neurons; organizations are networks of people; the global economy is a network of national economies, which are networks of markets, which in turn are networks of producers and consumers. Diseases and rumors both transmit themselves through social networks, and computer viruses propagate via the Internet.

This situation changed suddenly in 1998 with the publication of a paper in Nature by Duncan Watts and Steven Strogatz. In this paper the authors proposed a very simple model of a one-dimensional network on a ring (Watts & Strogatz, 1998).

Initially each node (*vertex*) in the network is only connected to its k nearest neighbors (k/2 on each side). K is called the degree of the network. Next, with a likelihood p, connections (*edges*) are chosen at random and connected to another vertex, also chosen randomly. With increasing p, more and more edges become randomly reconnected and finally for p = 1 all connections are random.

Thus, this simple model allows investigating the whole range from regular to random networks, including an intermediate range. The intermediate range proved to be crucial to the solution of the problem.

To show this, the authors introduced two measures: the clustering coefficient C, which is the likelihood that neighbors of a vertex will also be connected, and the path length L which is the average of the shortest distance between pairs of vertices counted in number of edges.

Watts and Strogatz showed that regular networks have a high C but also a very high L. In contrast, random networks have a low C and a low L. So, neither regular nor random networks explain the small-world phenomenon.

However, when p is only slightly higher than 0 (few edges randomly rewired) the path length L drops sharply, while the clustering coefficient hardly changes. *Thus networks with a small fraction of randomly rewired connections combine both high clustering and a small path length, and this is exactly the small-world phenomenon to be explained.* These networks were called *small-world* networks by the authors, who showed that such networks could be found in the nervous system, a social network of actors and the power grid of a nation. Also, they showed that small-world architecture might facilitate the spread of information (or infection) in networks.

8.3 Scale-Free

A second major discovery was made a year later by Barabasi and Albert (1999). They introduced the concept of *scale-free networks* and proposed a mechanism for their emergence. They proposed a model for the growth of a network where the likelihood that a newly added node will connect to a hub depends upon the degree of this hub. Thus, hubs that have a large number of nodes are more likely to get even more. Networks generated in this way are characterized by a degree distribution which can be described by a power law. In the case of the Barabasi Albert model the exponent is exactly 3. Networks with a power law degree distribution are called *scale-free*. It has been shown that many real networks in nature such as for instance the Internet, the World Wide Web, collaboration networks of scientists and networks of airports are likely to be scale-free. Scale-free networks have many interesting properties such as an extremely short path length.

The discovery of small-world networks in 1998 and of scale-free networks in 1999 was noted by scientists in many different fields, and set off a large body of theoretical and experimental research that is growing to this day. The definition of new types of networks was adding new perspectives to the previous known: ordered networks and random networks.

In retrospect these discoveries can be considered to be the starting point of the modern theory of networks. The field is so new that there are only few textbooks yet.

One of the most interesting and active research areas in modern network theory is the study of structure function relationships, in particular the relation between topological network characteristics and *synchronization dynamics* on these networks. The importance of the small world structure for the spread of infectious disease was already addressed in the original Watts and Strogatz paper

Barahona and Pecora used linear stability analysis and the master stability function (MSF) to study the synchronization in networks with complex topology (2002, 2008).

They showed that networks with a small-world topology may synchronize more easily than deterministic or fully random graphs, although the presence of small-world properties did not guarantee that the network will be synchronized.

Especially interesting results were obtained in the case of adding links to networks. First, in some cases adding links between two networks was shown to increase the synchronizability of the individual networks while decreasing the synchronizability of the combined network. Also, adding links to a single network could result in smaller path lengths but at the same time decreased synchronization. Of course this is reminiscent of the findings of Nishikawa et al. (2003).

A first important conclusion is that the modern theory of networks, which originated with the discovery of small-world and scale-free networks, is a very useful framework for the study of large scale networks (inside our bodies and outside, in society and the world). There are several reasons for this:

- the new theory provide us with powerful realistic models of complex networks;
- a large and still increasing number of measures becomes available to study topological and dynamical properties of these networks;
- this theory allows us to better understand the correlations between network structure and the processes taking place on these networks, in particular synchronization processes;

- by relating structure to function the notion of an optimal network (in terms of balancing segregation and integration, and performance and cost) can be formulated;
- the theory provides scenarios on how complex networks might develop, and how they might respond to different types of damage (random error versus targeted blocks, disruptions, attacks).

These considerations explain the motivation to apply modern network theory to neuroscience and to integrate it in Mind Force theory.

Some preliminary conclusions can be drawn from studies of functional connectivity in humans:

- most studies point in the direction of a small-world pattern for functional connectivity, although scale-free networks have also been described;
- the small-world topology of functional brain networks is very constant across techniques, conditions and frequency bands; tasks induce only minor modifications;
- the architecture of functional brain networks may reflect genetic factors and is related to cognitive performance;
- different types of brain disease can disrupt the optimal small-world pattern, sometimes giving rise to more random networks which may be associated with cognitive problems as well as a lower threshold for seizures (pathological hyper-synchronization).

8.4 Social Networks

The rich set of interactions between individuals in society results in complex community structure, capturing highly connected circles of friends, families or professional cliques in a social network. Thanks to frequent changes in the activity and communication patterns of individuals, the associated social and communication network is subject to constant evolution. Our knowledge of the mechanisms governing the underlying community dynamics is limited, but is essential for a deeper understanding of the development and self-optimization of society as a whole.

Palla, Barabasi & Wiczeck have developed an algorithm based on clique *percolation* (2007) that allows us to investigate the time dependence of overlapping communities on a large scale, and thus uncover basic relationships characterizing community evolution. Their focus is on networks capturing connections between families, friends, neighbors, the collaboration between scientists and the calls between mobile phone users. In scientific collaborations it seems that large groups persist for longer if they are capable of dynamically altering their membership, suggesting that an ability to change the group composition results in better adaptability. The behavior of small groups displays the opposite tendency – the condition for stability is that their composition remains unchanged. We also show that knowledge of the time commitment of members to a given community can be used for estimating the community's lifetime. These findings offer insight into the fundamental differences between the dynamics of small groups and large institutions.

The literature on crowd behavior and mob psychology (Freud, 1921, 1959; Canetti, 1962), marketing and advertising, can now be implemented in scientific terms. Even *memes*, once proposed by the evolutionary biologist Richard Dawkins (1976) as the psycho-cultural equivalents of genes, contagious ideas competing for survival and spreading just as epidemics, can become objects for a scientific study in this new framework. We have now a detailed and testable theory of social network dynamics. Rules governing human interactions and attractions, just as coupling between oscillators and the formation of networks, can be empirically studied and modeled. Also the somehow unexplained awkward moment of sudden simultaneous silence in a meeting or a cocktail party can find dynamical explanations. We are able to recognize the dynamical processes in which the spread of a geometric cluster percolates through a network, the way cascades can be triggered and flood the whole network or just be confined to a secluded area.

A visual example of how a community self-organizes is given by sociogram. A sociogram is a sociometric chart plotting the structure of interpersonal relations in a group situation, a kind of graph to represent social connections, and it was interesting to apply this technique to the scientific collaborations within a university institute in Rome.

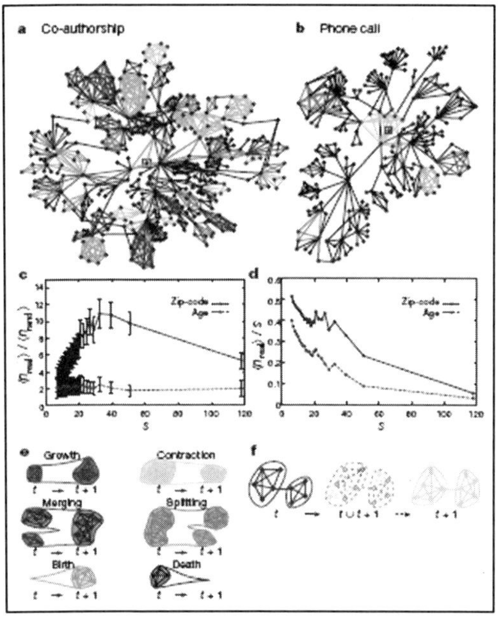

Fig. 47. Social network dynamics in collaborations (Palla, Barabasi &Vicsek, 2007).

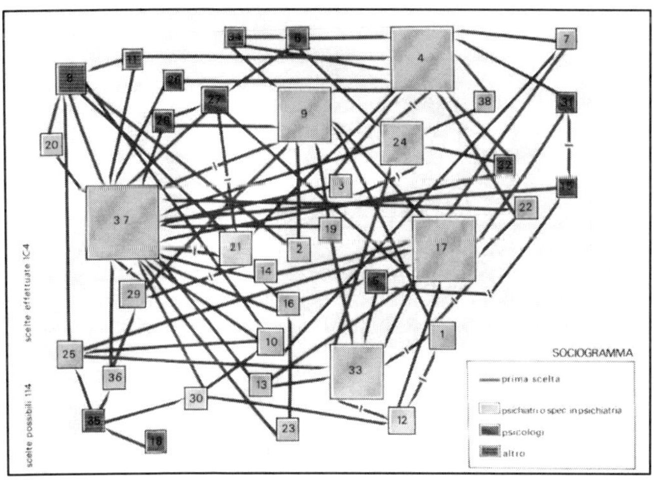

Fig. 48. Sociogram of scientific collaborations.

It is actually the first research institute we were working in, several years ago. The number of links and the dimension of nodes and hubs are clearly represented.

8.5 Conclusions

We have been exploring how the new construct of Mind Force can redefine the landscape of a vast amount of knowledge, bringing a new direction to the area of neuro and mind sciences.

Mind Force is based on the basic unity of four knowledge domains: complexity, networks, synchronization and field studies. These domains can, via their deep synergy, explain the attractions and cohesion of our biological, psychological and social identities. We hope this book will inspire new explorations, as in the words of Jorge Luis Borges:

> Time forks perpetually towards innumerable futures.
> I leave to the various futures (not to all) my garden of forking paths.

Appendix A

Manifesto

In September 2008, we organized a residential workshop in an ancient Charterhouse, at Pontignano, Siena, Italy. It was sponsored by the Department of Neurology and Behavioral Sciences, the Institute of Complex Systems Studies of the University of Siena and the Institute for Complexity Studies of Rome. During the conference, the panel produced the following Manifesto in order to sustain a shared public position.

The Mind Force Manifesto

Matter, mind, brain, body and society emerge from the same stream in the complexity of nature: we call this energy 'Mind Force'.

Connections of genes and molecules, neurons and hormones, thinking and language, people and organizations form a continuous flow of synchronized interactions.

Interactions between networks across many scales, from molecular, to biological, to cognitive and social, sustain its scaling and cascading. We recognize in this hyper network, the landscapes and fields of Mind Force.

The empowerment of individuals by Mind Force comes from integrating multiple perspectives and multiple complementary disciplines: simplicity and complexity are complimentary moments of our knowledge of each other, the world, and us.

Resources for research would follow the integrated path to a deeper knowledge based on multidisciplinary collaborations. Science is lauded by societies beyond the circle of experts and recognized as a natural mean of expanding lives, when it stays within the stream of Mind Force. This flow sustains the inner strength and growth of individuals and communities.

The first and key contributors to the Mind Force Manifesto are:

Alfredo Ancora
Tito Arecchi
Harald Atmanspacher
Peter Fonagy
Walter J. Freeman
Alessandro Giuliani
Guelfo Margherita
Lamberto Maffei
Chiara Mocenni
Franco Orsucci
Mario Reda
Sergio Rinaldi
Andrea Seganti
Steven Suomi
Giuseppe Vitiello
Joe Zbilut
Riccardo Zerbetto
Alberto Zucconi

Appendix B

Modeling tools

A pair of oscillators interacting through phase differences satisfies equations of the form

$$\theta_1' = \omega_1 + H_1(\theta_2 - \theta_1),$$
$$\theta_2' = \omega_2 + H_2(\theta_1 - \theta_2),$$

Here θ_i are the phases of the oscillators, ω_i are the frequencies of the uncoupled oscillators, and H_i are smooth 2π-periodic functions of the phase differences.

From a more abstract viewpoint, synchronization has been identified as a generic form of collective behavior in ensembles of dynamical systems with long range coupling. Several models that capture the essence of synchronization phenomena have been thoroughly studied over the last few decades. Kuramoto (1984) for instance, has analyzed an ensemble of N coupled phase oscillators, governed by the equations

$$\dot{\phi}_i(t) = \Omega_i + \frac{\varepsilon}{N}\sum_{j=1}^{N} \sin(\phi_j - \phi_i),$$

$i = 1,\ldots,N$, where $\varepsilon > 0$ is the strength of coupling. In the absence of coupling, $\varepsilon = 0$, each oscillator i performs a uniform angular motion with its natural frequency Ω_i. For $\varepsilon \neq 0$, the oscillators are globally coupled in the sense that the strength of the pair interaction does not depend on their relative position, but only on their relative state. In other words, each oscillator interacts with the rest of the system through global averages only.

Kuramoto has shown that, in the limit $N \to \infty$, there exists a critical value ε_c of the coupling intensity such that, for $\varepsilon > \varepsilon_c$, a subensemble of oscillators becomes entrained in periodic orbits with the same frequency, whereas the other oscillators remain unsynchronized.

A great deal of attention has been paid to the synchronization of ensembles formed by identical elements, especially, in the case where the individual dynamics is chaotic. Both continuous and discrete-time dynamics have been considered. Kaneko (1994) has introduced globally coupled chaotic maps as a mean-field model of lattice maps, which are extensively used to model complex extended systems (Kaneko, 1993).

For an ensemble of N maps whose individual dynamics is governed by the equation $w(t+1) = f[w(t)]$, global coupling is introduced as

$$\begin{aligned} w_i(t+1) &= (1-\varepsilon)\, f[w_i(t)] + \frac{\varepsilon}{N}\sum_{j=1}^{N} f[w_j(t)] \\ &= (1-\varepsilon)\, f[w_i(t)] + \varepsilon\, \overline{f(w)}, \end{aligned}$$

$i = 1,\ldots,N$, with $\varepsilon \in [0, 1]$. While for $\varepsilon = 0$ the elements evolve independently, for $\varepsilon = 1$ they become fully synchronized after the first time step.

Full synchronization is understood here as a situation where the individual states of all the elements in the ensemble coincide, i.e. where the trajectory of the system in phase space is restricted to the subspace $w_1 = w_2 = \cdots = w_N$. In this situation, the evolution of all the elements coincides with that of an independent element. The state of full synchronization can be asymptotically approached as the system evolves even for $\varepsilon < 1$. It has been shown that, if the individual dynamics is chaotic, full synchronization is linearly stable for $\varepsilon > \varepsilon_c$, where the critical value ε_c is related to the maximal Lyapunov exponent λ_M of the individual dynamics, as $\varepsilon_c = 1 - \exp(-\lambda_M)$. For nonchaotic individual dynamics where $\lambda_M < 0$, full synchronization is a stable state for any $\varepsilon > 0$. The connection between ε_c and λ_M makes it clear that the transition to full synchronization in chaotic systems, which has the character of a critical phenomenon, results from the competition between the stabilizing effect of global coupling and the inherent instability of chaotic orbits. Note carefully that the critical value ε_c does not depend on N, so that the synchronization threshold is the same for any size of the coupled ensemble.

For coupling strengths just below ε_c the system evolves asymptotically to a state of partial synchronization in the form of *clustering*, where the elements become divided into groups (Kaneko, 1989). Within each cluster the elements are fully synchronized but different clusters have different trajectories.

For large systems, the dynamics in the clustering regime is highly multistable and exhibits glassy-like features (Crisanti et al., 1996; Manrubia & Mikhailov, 2001). In contrast with the critical value ε_c, the stability properties of the clustering regime are strongly dependent on the system size (Abramson, 2000).

Cross-Coupled Extended Systems provide a versatile collection of models for a wide class of complex natural phenomena, ranging from pattern formation in physicochemical reactions, to biological morphogenesis, to evolutionary processes.

It is therefore interesting to consider how these systems behave under the effect of mutual interactions and, in particular, study the synchronization properties of their coevolution when they are mutually coupled by algorithms similar to the scheme of coupling already presented (Zanette & Morelli, 2003).

P.M. Gleiser and D.H. Zanette (2006) analyzed the interplay of synchronization and structure evolution in an evolving network of phase oscillators. An initially random network is adaptively rewired according to the dynamical coherence of the oscillators, in order to enhance their mutual synchronization. They showed that the evolving network reaches a small-world structure. Its clustering coefficient attains a maximum for an intermediate intensity of the coupling between oscillators, where a rich diversity of synchronized oscillator groups is observed. In the stationary state, these synchronized groups are directly associated with network clusters.

Their model consists of an ensemble of N coupled phase oscillators, whose individual evolution is given by

$$\dot{\phi}_i = \omega_i + \frac{r}{M_i} \sum_{j=1}^{N} W_{ij} \sin(\phi_j - \phi_i),$$

$i = 1,\ldots,N,$ where ω_i is the natural frequency of oscillator I and r is the coupling strength. The weights W_{ij} define the adjacency matrix of the interaction network: $W_{ij} = 1$ if oscillator i interacts with oscillator j, and 0 otherwise. The number of neighbors of oscillator i is $M_{ij} = \sum_j W_{ij}$. The adjacent matrix is symmetric, $W_{ij} = W_{ji}$, so that the network is a non directed graph.

Bibliography

Abraham, F. D., Abraham, R. & Shaw, C. D. (1990). *A Visual Introduction To Dynamical Systems Theory For Psychology.* Santa Cruz, CA: Aerial Press.

Abramson, G. (2000). Long transients and cluster size in globally coupled maps, *Europhys. Lett.* 52(6), 615-619.

Aiello, L. C., & Wheeler, P. (1995). The expensive-tissue hypothesis, *Current Anthropology* 36:199-221.

Ainsworth, M. D. (1969). Object relations, dependency, and attachment: a theoretical review of the infant-mother relationship, *Child Dev.* 40, 969-1025.

Ainsworth, M. D. (1985). Attachments across the life span. *Bull. N.Y. Acad. Med.* 61, 792-812.

Albert, R. and Barabasi, A. L. (2002). Statistical mechanics of complex networks, *Reviews of Modern Physics,* Volume 74.

Ambrose, S. H. (2001). Paleolithic technology and human evolution, *Science* 291, 1748-1753.

Amit, D. J. (1989). *Modeling Brain Function: The World Of Attractor Neural Networks.* New York, NY: Cambridge University Press.

Andrew, T. J. & Purves, D. (2005). The wagon wheel illusion in continuous light, *Trends in Cog Neurosci.* 9(6): 261-263.

Aoki, K. & Feldman M. V. (1987). Toward a theory for the evolution of cultural, *Proceedings of the National Academy of Sciences USA,* 84:7, 164-68.

Arbib, M. A. & Mundhenk, T. N. (2005). Schizophrenia and the mirror system: an essay,*Neuropsychologia* 43, 268-280.

Arbib, M. A. (2001). Co-evolution of human consciousness and language, *Ann. N.Y. Acad. Sci.* 929, 195-220.

Arbib, M. A. (2002). The mirror system, imitation, and the evolution of language, in *Imitation in Animals and Artifacts* (Chrystopher Nehaniv and Kerstin Dautenhahn, Editors), The MIT Press.

Arbib, M. A. (2005). From monkey-like action recognition to human language: an evolutionary framework for neurolinguistics, *Behav. Brain Sci.* 28, 105-124.

Arbib, M. A. (2007). Other faces in the mirror: a perspective on schizophrenia, *World Psychiatry* 6, 11-14.

Ashby, W. R. (1952). *Design for a Brain.* London: Chapman and Hall.

Atmanspacher, H. (2004). Quantum theory and consciousness: an overview with selected examples, *Discrete Dynamics in Nature and Society* 1, 51-73.

Austin, J. H. (2006). *Zen-Brain Reflections*. Cambridge: MIT Press.

Baer, R. A. (2003). Mindfulness training as a clinical intervention: a conceptual and empirical review, *Clinical Psychology: Science and Practice* 10, 125-143.

Bak, P., Tang, C. & Wiesenfeld, K. (1984). Self-organized criticality: an explanation of 1/f noise, *Physical Review Letters* 59, 381-384.

Bak, P. (1990). Simulation of self-organized criticality, *Physica Scripta,* T33.

Bak P. (1996). *How Nature Works, The Science Of Self-Organized Criticality*. New York, NY, USA: Copernicus.

Barabasi, A.-L. & Albert, R. (1999). Emergence of scaling in random networks, *Science* 286, 509-512.

Barabasi, A.-L. (2005). Taming complexity, *Nature* 1: 68-70.

Barahona, M. & Pecora, L. M. (2002). Synchronization in small-world systems, *Physical Review Letters* 89, 054101.

Bartels, A. & Zeki, S. (2000). The neural basis of romantic love, *Neuroreport* 11(17):3829-3834.

Bates, E. (1979). *The Emergence Of Symbols, Cognition And Communication In Infancy*. New York: Academic Press.

Bates, E. & MacWhinney, B. (1987). Competition, variation and language learning. MacWhinney, B. (ed.). *Mechanisms of Language Acquisition* (pp.157-193), Hillsdale, NJ: Erlbaum.

Bateson, G. (1987). *Steps To An Ecology Of Mind, Collected Essays In Anthropology, Psychiatry, Evolution, And Epistemology*. Northvale, N.J: Aronson.

Bergson, H. B. (1911). *Creative Evolution,* tr. Arthur Mitchell, Henry Holt and Company

Bion, W. R. (1992). *Cogitations*. London: Karnac Books.

Boccaletti, S., Grebogi, C., Lai, Y.-C., Mancini, H. & Maza, D. (2000). The control of chaos: theory and applications, *Physics Reports* 329[3], 103-197.

Boccaletti, S., Pecora, L. M., & Pelaez, A. (2001). Unifying framework for synchronization of coupled dynamical systems, *Phys. Rev. E.* 63, 066219.

Bodmer, W. F. & Cavalli-Sforza, L. L. (1976). *Genetics, Evolution And Man*. San Francisco: Freeman.

Bohm, D. & Hiley, B. J. (2008). *The Undivided Universe*. London: Routledge.

Borges J. L. (1964). *Labyrinths, Selected Stories & Other Writings*. New York: New Directions Pub. Corp.

Bowlby, J. (1978). Attachment theory and its therapeutic implications, *Adolesc. Psychiatry* 6, 5-33.

Bowlby, J. (1982). Attachment and loss: retrospect and prospect, *Am. J. Orthopsychiatry* 52, 664-678.

Boyd, R. & Richerson, P. J. (1985). *Culture And The Evolutionary Process*. Chicago: University of Chicago Press.

Branigan, H. P., Pickering, M. J. & Cleland, A. A. (2000). Syntactic co-ordination in dialogue, *Cognition B* 75, 13-25.

Breuer, J. & Freud, S. (1895). *Studien über Hysterie.* Leipzig: F. Deuticke.

Cancedda, L., Putignano, E, Sale, A., Viegi, A., Berardi, N. & Maffei, L. (2004). Acceleration of visual system development by environmental enrichment, *J. Neurosci.* 24: 4840-4848.

Canetti, E. (1962). *Crowds And Power.* London: Gollancz.

Carbonara, D. (ed.) (2005). *Technology Literacy Applications in Learning Environments.* Hersey, NJ: Idea Group.

Carroll, T. L. & Pecora, L. M. (1993a). Stochastic resonance and crises, *Physical Review Letters* 70, 576-579.

Carroll, T. L. & Pecora, L. M. (1993b). Using chaos to keep period-multiplied systems in phase, *Physical Review E* 48, 2426-2436.

Chalmers, D. J. (1995). *Explaining Consciousness: The Hard Problem.* MIT Press.

Chalmers, D. J. (1996). *The Conscious Mind. In Search of a Fundamental Theory.* Oxford: Oxford University Press.

Chan, P.A. & Rabinowitz, T. (2006). A cross-sectional analysis of video games and attention deficit hyperactivity disorder symptoms in adolescents, *Ann Gen Psychiatry,* Oct. 24, 5:16.

Changeux, J.-P. (2004). *The Physiology of Truth: Neuroscience and Human Knowledge.* Harvard University Press.

Chomsky, N. (1975). *Reflections on Language.* New York: Pantheon,

Chomsky, N. (1993). *Language And Thought* (1st ed ed.), Wakefield, R.I: Moyer Bell.

Churchland, P. M. (1995). The puzzle of conscious experience, *Scientific American* 273, 80-86.

Corrao, F. (1986). Il concetto di campo come modello teorico, in *Gruppo e Funzione analitica,* VII, N.1.

Corrao, F. (1992). *Modelli Psicoanalitici. Mito, Passione, Memoria.* Bari: Editori Laterza.

Crick, F. & Koch, C. (2003). A framework for consciousness, *Nature Neuroscience* 6-2.

Crutchfield, J. P. & Kaneko, K. (1987). Phenomenology of Spatiotemporal Chaos, in *Directions in Chaos,* World Scientific, 272-353.

Dale, R. & Spivey, M. J. (2006). Unravelling the dyad: Using recurrence analysis to explore patterns of syntactic coordination between children and caregivers in conversation, *Language Learning* 56, 391-430.

Damasio, A. R. (1998a). Emotion in the perspective of an integrated nervous system, *Brain Res. Brain Res. Rev.* 26, 83-86.

Damasio, A. R. (1998b). Investigating the biology of consciousness, *Philos. Trans. R. Soc. Lond. B Biol. Sci.* 353, 1879-1882.

Damasio, A. R. (1999). How the brain creates the mind, *Sci. Am.* 281, 112-117.

Darwin, C. R. (1909). *The Origin Of Species.* New York: P.F. Collier & Son.

Dawkins, R. (1976). *The Selfish Gene.* Oxford: Oxford University Press.

Dawkins, R. (1982). *The Extended Phenotype.* San Francisco, CA: Freeman.

De Risio, S. & Orsucci, F. (2004). *Bioethics In Complexity, Foundations And Evolutions.* London: Imperial College Press.

Dennett, D. C. (1991). *Consciousness Explained.* Boston: Little, Brown and Co.

Descartes, R. (1988). *The Philosophical Writings Of Descartes.* Cambridge: Cambridge University Press.

Donald, M. (1997). Origins of the modern mind: Three stages in the evolution of culture and cognition, *Behavioral and Brain Sciences* 16 (4): 737-791.

Dubé, J. & Desprès, P. (2000). The control of dynamical systems – recovering order from chaos, in Y. Itikawa (Ed.), *The Physics of Electronic and Atomic Collisions.* Woodbury, N.Y.: AIP.

Eccles, J. C. (1990). A unitary hypothesis of mind–brain interaction in cerebral cortex. *Proc. R. Soc. Lond. B* 240, 433–451.

Eckmann, J.-P., Kamphorst, S. O. & Ruelle, D. (1987). Recurrence plots of dynamical systems, *Europhysics Letters* 5, 973-977.

Eco, U. (1989). *The Open Work.* Cambridge, Mass: Harvard University Press.

Eco, U. (1990). *The Limits of Interpretation.* Bloomington, Ind: Indiana University Press.

Edelman, G. M. (1989). *Neural Darwinism: The Theory of Neuronal Group Selection.* New York: Basic Books.

Edelman, G. M. & Tononi, G. (2000). *A Universe Of Consciousness. How Matter Becomes Imagination.* New York, NY: Basic Books.

Edwards, J. C. W. (2005). Is Consciousness Only a Property of Individual Cells? *Journal of Consciousness Studies* 12.

Eliasmith, C. (2005). A unified approach to building and controlling spiking attractor networks, *Neural Computation* 17(6): 1276-1314.

Ellenberger, H. F. (1970). *The Discovery Of The Unconscious. The History And Evolution Of Dynamic Psychiatry.* New York: Basic Books.

Beck F. & Eccles J. (1992). Quantum Aspects of Brain Activty and the Role of Consciousness, *Proc. Nat. Acad. Sci. USA* 89: 11357-11361.

Feldman, M. W. & Laland, K. N. (1996). Gene-culture coevolutionary theory, *Trends in Ecology and Evolution* 11:453-457.

Fisher, H. E., Aron, A., Mashek, D., Li, H. & Brown, L. L. (2002). Defining the brain systems of lust, romantic attraction, and attachment – *Archives of sexual behavior*, Vol. 31, No. 5.

Fonagy, P. & Target, M. (1997). Attachment and reflective function: Their role in self-organization, *Development and Psychopathology* 9, 679-700.

Fonagy, P. (2002). *Affect Regulation, Mentalization, And The Development Of The Self.* New York: Other Press.

Freeman, W. J. (1975). *Mass Action In The Nervous System.* NY, Academic Press.

Freeman, W. J. (1978). Spatial properties of an EEG event in the olfactory bulb and cortex. *Electroencephalogr. Clin. Neurophysiol.* 44, 586-605.

Freeman, W. J. (1997). Neurohumoral brain dynamics of social group formation. Implications for autism, *Ann. N.Y. Acad. Sci.* 807, 501-503.

Freeman, W. J. (1999). *HOW BRAINS MAKE UP THEIR MINDS.* London: Weidenfeld & Nicolson.

Freeman, W. J. (2000). Mesoscopic neurodynamics: from neuron to brain, *J. Physiol. Paris* 94, 303-322.

Freeman, W. J. (2006). A cinematographic hypothesis of cortical dynamics in perception. *Int. J. Psychophysiol.* 60, 149-161.

Freeman, W. J. (2007). Indirect biological measures of consciousness from field studies of brains as dynamical systems. *Neural Netw.* 20, 1021-1031.

Freeman, W. J., & Vitiello, G. (2006). Nonlinear brain dynamics as macroscopic manifestation of underlying many-body field dynamics. *Physics of Life Reviews* 3, 93–118.

Freud S, (1959). *Collected Papers*. New York: Basic Books.

Freud, S. (1985). *The Complete Letters Of Sigmund Freud To Wilhelm Fliess,* 1887-1904, Cambridge, Mass: Belknap Press.

Gaddini, E. (1969). On imitation, *Int. J. Psycho-Anal.* 50:475-484.

Gaddini, E., (1989). *Scritti, 1953-1985*. Milano: R. Cortina.

Gazzaniga, M. S. (2005). *The Ethical Brain.* New York: The Dana Press.

Gibson, J. J. (1979). *The Ecological Approach To Visual Perception.* Boston: Houghton Mifflin.

Gleiser, P. M. & Zanette, D. H. (2006). Synchronization and structure in an adaptive oscillator network, *Eur. Phys. J. B* 53, 233–238

Goleman, D. (1977). *The Varieties Of The Meditative Experience.* New York: Irvington Publishers.

Goleman, D. (1995). *Emotional Intelligence.* New York: Bantam Books.

Green, C.S. & Bavelier, D. (2003). Action video game modifies visual selective attention. *Nature,* Vol. 423, 534-537.

Groddeck, G. & Freud, S. (1970). *Der Mensch und sein Es. Briefe, Aufs tue. Biografisches*, Wiesbaden: Limes Verlag.

Grush, R. & Churchland, P. (1995). Gap's in Penrose's Toilings, *Journal of Consciousness Studies* 2(1), pp. 10-29.

Hagan, S., Hameroff, S. & Tuszynski, J. (2002). Quantum computation in brain microtubules? Decoherence and biological feasibility, *Physical Reviews E* 65.

Haken, H. & Stadler, M. (1990). *Synergetics Of Cognition*, Berlin: Springer-Verlag.

Haken, H. & Mikhailov, A. S. (1993). *Interdisciplinary Approaches to Nonlinear Complex Systems.* Berlin: Springer-Verlag.

Hall, E. T. (1983). *The Dance of Life: The Other Dimension of Time.* New York: Anchor Press/Doubleday.

Hameroff, S. (1987). *Ultimate Computing.* Elsevier.

Hameroff, S. (2006). Consciousness, neurobiology and quantum mechanics, in: *The Emerging Physics of Consciousness* (Ed. Tuszynski, J.), NY: Elsevier.

Hebb, D. O. (1949). *The Organization Of Behavior.* New York, NY: Wiley.

Hofer, M. A. (1984). Relationships as regulators: a psychobiologic perspective on bereavement, *Psychosom. Med.* 46, 183-197.

Hofer, M. A. (1994a). Early relationships as regulators of infant physiology and behavior. *Acta Paediatr. Suppl.* 397, 9-18.

Hofer, M. A. (1994b). Hidden regulators in attachment, separation, and loss, *Monogr. Soc. Res. Child Dev.* 59, 192-207.

Hofstadter, D. (2007). *I Am A Strange Loop*. New York: Basic Books.

Hopfield, J. J. (1982). Neural networks and physical systems with emergent collective computational abilities, *Proceedings of the National Academy of Sciences* 79, 2554–2558.

Hume, D. (1854). *The Philosophical Works Of David Hume*. Boston: Little, Brown and Company.

Husserl, E. (1980). *Collected Works*. Boston: The Hague

Insel, T. R., Young, L. & Wang, Z. (1997). Molecular aspects of monogamy, *Ann. N Y Acad. Sci.* Jan 15; 807:302-16.

Jacob, S. & McClintock, M. K. (2000). Psychological state and mood effects of steroidal chemosignals in women and men, *Horm.Behav.* 37, 57-78.

James, W. (1967). *The Writings Of William James* (McDermott JJ ed), New York, Random House.

Jones, F. J. (1995). *The Structure of Petrarch's Canzoniere*. Cambridge: Brewer.

Jourdain, R. (1997). *Music, The Brain, And Ecstasy: How Music Captures Our Imagination*. New York: Harper Perennial.

Jung, C. G. & Pauli, W. (1955). *The Interpretation of Nature and the Psyche*. NY: Pantheon Books.

Kandel, E. R. (1999). Biology and the Future of Psychoanalysis: A New Intellectual Framework for Psychiatry Revisited, *Am. J. Psychiatry* 156: 505-524.

Kaneko, K. (1993). *Theory And Applications Of Coupled Map Lattices*. Chichester: Wiley.

Kaneko, K. (1994). Relevance of clustering to biological networks, *Physica D* 75(1-3), 55-73.

Kauffman, S. A. (1993). *The Origins Of Order, Self Organization And Selection In Evolution*. New York: Oxford University Press.

Kauffman, S. A. (1995). *At Home In The Universe. The Search For Laws Of Self-Organization And Complexity*. New York: Oxford University Press.

Kauffman, S. A. (2000). *Investigations*. Oxford: Oxford University Press.

Kelso, J. A. S. & Clark, J. E. (1982). *The Development Of Movement Control And Coordination*. Chichester, West Sussex: Wiley.

Kelso, J. A. S., Mandell, A. J., Shlesinger, M. F. & Haken, H. (1988). *Dynamic Patterns In Complex Systems*. Singapore: World Scientific.

Kelso, J. A. S. (1995). *Dynamic Patterns, The Self-Organization Of Brain And Behavior*. Cambridge, Mass: MIT Press.

Kernberg, P.F. (2007). *Beyond the Reflection: The Role of the Mirror Paradigm in Clinical Practice*. New York: Other Press.

Kornhuber, H. H. & Deecke, L. (1965). Hirnpotentialänderungen bei Willkürbewegungen und passiven Bewegungen des Menschen: Bereitschaftspotential und Reafferente Potentiale, *Pflügers Arch. Ges. Physiol.* 284, 1-17.

Kozma, R. & Freeman, W. J. (2003). Basic principles of the KIV model and its application to the navigation problem, *J. Integr. Neurosci.* 2, 125-145.

Kozma, R., Puljic, M., Balister, P., Bollobas, B. & Freeman, W. J. (2005). Phase transitions in the neuropercolation model of neural populations with mixed local and non-local interactions, *Biol. Cybern.* 92, 367-379.

Kupfer, D. J., Monk, T. H. & Barchas, J. D. (1988). *Biological Rhythms And Mental Disorders.* New York: Guilford Press.

Lacan, J. (1937). *Ecrits.* New York: W. W. Norton (New edition 2005).

LaForge, R. (2004). The early development of the interpersonal system of personality (ISP), *Multivariate Behavioral Research,* Volume 39, Issue 2, pages 359-378.

Laland, K. N., Odling-Smee, J. & Feldman, M. W. (2000). Niche construction, biological evolution, and cultural change, *Behavioral And Brain Sciences* 23, 131–175.

LeDoux, J. E. (2002). *Synaptic Self, How Our Brains Become Who We Are.* New York: Viking.

Leibniz, G. W. (1996). *New Essays On Human Understanding.* Cambridge: Cambridge University Press.

Lewin, K. (May, 1939). Field theory and experiment in social psychology: concepts and methods, *The American Journal of Sociology*, Vol. 44, No. 6, pp. 868-896.

Lewin, K. (1943). Psychological ecology, in: D. Cartwright (Ed.) (1952), *Field Theory in Social Science, Social Science,* Paperbacks, London.

Lewontin, R. C. (1983). Gene, organism, and environment, in: *Evolution From Molecules To Men,* ed. D. S. Bendall. Cambridge University Press.

Libet, B. (1993). *Neurophysiology Of Consciousness: Selected Papers And New Essays.* Boston: Birkhauser.

Libet, B. (1994). A testable field theory of the mind-brain interaction, *J. Consciousness Stud.* 1, 119-126.

Libet, B. J. (1996). Conscious mind as a field, *J. Theor. Biol.* 178, 223-2243.

Libet, B. J. (1997). Conscious mind as a force field: a reply to Lindahl & Arhem, *J. Theor. Biol.* 185, 137-138.

Libet, B. J., Freeman, A. & Sutherland, K. (1999). *The Volitional Brain, Towards A Neuroscience Of Free Will.* Thorverton, UK: Imprint Academic.

Libet, B. J. (2006). Reflections on the interaction of the mind and brain, *Progress in Neurobiology* 78, 322–326.

Lieberman, P. (1991). *Uniquely Human: The Evolution of Speech, Thought, and Selfless Behavior.* Harvard University Press, Cambridge, Massachusetts

Lindahl, B. I. & Arhem, P. (1994). Mind as a force field: comments on a new interactionistic hypothesis, *J. Theor. Biol.* 171, 111-122.

Lindahl, B.I. and Arhem, P. (1996). The mental force field hypothesis: a reply to Libet. *J. Theor. Biol.* 178, 225-226.

Lorenz, E. N. (1993). *The Essence Of Chaos.* Seattle: University of Washington Press.

Lorenzana, Alvarez de, J. M. & Ward, L. M. (1987). On evolutionary systems, *Behav. Sci.* Jan.; 32(1):19-33.

Margenau, H. (1984). *The Miracle of Existence.* Woodbridge, Conn.: Ox Bow Press.

Marwan, N. (2003). *Encounters With Neighbours,* doctoral dissertation, University of Potsdam.

Mathiak, K. & Weber, R. (2006). Toward brain correlates of natural behavior: fMRI during violent video games, *Hum Brain Mapp.* Dec.;2 7(12):948-956.

Maturana, H. R. & Varela, F. J. (1980). *Autopoiesis And Cognition, The Realization Of The Living.* Dordrecht, Holland: D. Reidel Pub. Co.

Maturana, H. (2002). Autopoiesis, Structural Coupling and Cognition: A history of these and other notions in the biology of cognition, *Cybernetics & Human Knowing* 9 (3-4): 5-34.

McClintock, M. K., Bullivant, S., Jacob, S., Spencer, N., Zelano, B. & Ober, C. (2005). Human body scents: conscious perceptions and biological effects, *Chem. Senses* 30, Suppl. 1, i135-i137.

Mittag, M. & Volker, W. (2003). The circadian clock of the unicellular eukaryotic model organism Chlamydomonas reinhardtii, *Biological Chemistry,* vol. 384, no. 5, pp. 689-695.

Neri C. (1993). Campo. *Psiche,* Anno 1, n. 2, 1993, 289-295.

Neri, C. (1998). *Group.* London: Jessica Kingsley.

Nishikawa, T., Motter, A. E., Lai, Y.C. & Hoppensteadt, F. C. (2003). Heterogeneity in oscillator networks: are smaller worlds easier to synchronize, *Physical Review Letters.*

Norman, D. A. (1988). *The Design of Everyday Things.* New York: Doubleday

Norman, D. A. (1999). Affordances, conventions, and design, *Interactions* 6 (3), p. 38-41.

Orsucci, F. (1996a). Chaos and complexity in psychopathology. *Giorn. Ital. Psicopat.* 3.

Orsucci, F. (1996b). Behcet's disease and psychosomatic patterns of thinking. A single case study, *Psychother. Psychosom.* 65, 112-114.

Orsucci, F., (ed) (1998). *The Complex Matters Of The Mind.* Singapore: World Scientific.

Orsucci, F., Walters, K., Giuliani, A., Webber, C., Jr. & Zbilut, J. (1999). Orthographic structuring of human speech and texts, *International Journal of Chaos Theory and Applications,* Vol. 4-2, pp. 80-88.

Orsucci, F. (2000). *Olfaction: New Frontiers for Cognitive Sciences* (Edited by C. Di Natale, A. D'Amico & P. Siciliano), *Sensors and Microsystems.* Singapore: World Scientific Publishing.

Orsucci, F. (2002a). Linguistic landscapes, *Chaos & Complexity Letters* 1, 2.

Orsucci, F. (2002b). *Changing Mind.* Singapore: World Scientific.

Orsucci, F., Giuliani, A. & Zbilut. J. (2004). Structure & Coupling of Semiotic Sets, *Experimental Chaos*, American Institute of Physics, AIP Proceedings, Vol. 742, pp. 83-93.

Orsucci, F. & Sala, N. (2005). Virtual reality, telemedicine and beyond (Carbonara, D., ed.), *Technology Literacy Applications in Learning Environments* (pp. 349-357), Idea Group, Hersey.

Orsucci, F., Giuliani, A., Webber, C., Zbilut, J., Fonagy, P. & Mazza, M., (2006). Combinatorics & synchronization in natural semiotics, *Physica A: Statistical Mechanics and its Applications* – Elsevier 361: 665–676.

Orsucci, F. & Sala, N. (2006). *New Research On Chaos And Complexity*. New York: Nova Science Publishers.

Orsucci, F. (2006). The paradigm of complexity in clinical neuro-cognitive science, *The Neuroscientist* – SAGE, Vol. 12- 4, 1-10.

Orsucci, F. (2009). Mind Force, modeling a biophysical hyperstructure, *Chaos & Complexity Letters,* 4-1.

Ott, E. Grebogi, C. & Yorke, J. A. (1990). Controlling chaos, *Phvsical Review Letters*, vol. 64. no. 11, pp. 1196-1199.

Paetkau, V., Roderic, E. & Reinhard, I. (2006). A model for generating circadian rhythm by coupling ultradian oscillators, *Theoretical Biology and Medical Modelling* 3:12.

Paley, B. (2002). *TextArc* – extracting new meaning and relationships from literary works – new visual cataloguing approach, http://www.textarc.org/

Palla, G., Barabasi, A.-L. & Vicsek, T. (2007). Quantifying social group evolution, *Nature*, Vol. 446.

Panksepp, J. (1998). *Affective Neuroscience, The Foundations Of Human And Animal Emotions*. New York: Oxford University Press.

Panksepp, J. (2004). *Textbook Of Biological Psychiatry*. Hoboken, N.J: Wiley-Liss.

Pecora, L. M. & Carroll, T. L. (1990). Synchronization in chaotic systems, *Physical Review Letters* 64, 821-824.

Pecora, L. M. & Carroll, T. L. (1991). Driving systems with chaotic signals, *Physical Review A* 44, 2374-2383.

Pecora, L. M. & Barahona, M. (2008). Synchronization of oscillators in complex networks, *Pramana*, Vol. 70, No. 6.

Peirce, C. S. (1931). *Collected papers of Charles Sanders Peirce*. Cambridge: Harvard University Press.

Penrose, R. (1989). *The Emperor's New Mind*. Oxford: Oxford University Press.

Penrose, R. (1994). *Shadows of the Mind*. Oxford: Oxford University Press.

Petitot, J. (1999). *Naturalizing Phenomenology, Issues In Contemporary Phenomenology And Cognitive Science*. Stanford CA: Stanford University Press.

Piaget, J. (1930). *The Child's Conception of Physical Causality*. London: Kegan Paul, Trench, & Trubner (Original work published in French in 1927).

Popper, K. R. (1993). A discussion on the mind-body problem, *Theor. Med.* 14, 167-180.

Porphyry of Tyre (1940). *Isagoge* in Catalogus Codicum astrologorum Graecorum, Stefan Weinstock, ed. Franz Cumon, ed. (Brussels): V.4, 187-228.

Port, R., Tajima, K. & Cummins, F. (1999). *Non-linear Analysis of Developmental Processes* (van der Maas and van Geert ed), Amsterdam, NE: Elsevier.

Pribram, K. (1971). *Languages of the Brain*. Prentice-Hall, New Jersey,.

Purves, D. (2003). *Why We See What We Do: An Empirical Theory of Vision*. Sinauer Associates, Inc.

Pyragas, K. (1996). Weak and strong synchronization of chaos. *Physical Review E* 54, R4508-R4511.

Pyragas, K. (2001). Control of chaos via an unstable delayed feedback controller. *Phys. Rev. Lett.* 86, 2265-2268.

Quillian, M. R. (1967). Word concepts: a theory and simulation of some basic semantic capabilities, *Behavioral Science* 12: 410-430.

Quintero, J. E., Kuhlman, S. J. & McMahon, D. G. (2003). The biological clock nucleus: a multiphasic oscillator network regulated by light, *The Journal of Neuroscience,* September 3, 23(22):8070-8076.

Reppert, S. M. & Weaver, D. R. (2002). Coordination of circadian timing in mammals, *Nature* 418, 6901, pp. 935-941.

Ricciardi, L. M. & Umezawa, H. (1967). Brain and physics of many-body problems. *Kybernetik* 4, 44-48.

Richards, G. (1997). *Race, Racism, And Psychology: Towards A Reflexive History*. New York: Routledge.

Richardson, D. C., & Dale, R. (2005). Looking to understand: The coupling between speakers' and listeners' eye movements and its relationship to discourse comprehension, *Cognitive Science* 29, 39-54.

Rinaldi, S. (1998b). Laura and Petrarch: An intriguing case of cyclical love dynamics, *SIAM Journal on Applied Mathematics* 58, 1205-1221.

Rizzolatti, G. & Arbib, M. A. (1998). Language within our grasp, *Trends in Neuroscience* 21, 188-194. Vintage Books ed. (rev. and expanded ed.) New York: Vintage Books.

Rossi, E. L. (1996). The psychobiology of mind-body communication: The complex, self-organizing field of information transduction, *Bio. Systems* 38, 199-206.

Sacks, O. W. (2004). In *the River of Consciousness*. The New York Review of Books: volume 51, number 1.

Sacks, O. W. (2007). *Musicophilia, Tales Of Music And The Brain*. New York: Alfred, A. Knopf.

Sale, A., Putignano, E., Cancedda, L., Landia, S., Cirulli, F., Berardi, N. & Maffei, L. (2004). Enriched environment and acceleration of visual system development, *Neuropharmacology* 47, 5, 649-660.

Schopen, A. & Freeman, B. (1992). Meditation: the forgotten western tradition, *Counseling and Values*, vol. 36, no. 2, pp. 123-34.

Shockley, K., Santana, M. V. & Fowler, C. A. (2003). Mutual interpersonal postural constraints are involved in cooperative conversation, *Journal of Experimental Psychology: Human Perception and Performance* 29, 326-332.

Siegel, D. J. (2007). *The Mindful Brain*. New York: Norton.

Skarda, C. A. & Freeman, W. J. (1987). How brains make chaos in order to make sense of the world, *Behavioral and Brain Sciences* 2, 10.

Somers, D. & Kopell, N. (1995). Waves and synchrony in networks of oscillators of relaxation and non-relaxation type, *Physica D* 89, 169-183.

Sperry, R. W. (1947). Cerebral regulation of motor coordination in monkeys following multiple transections of sensorimotor cortex, *J. Neurophysiol.* 10, 275-294.

Sperry, R. W. (1980). Mind-brain interaction: mentalism, yes; dualism, no., *Neuroscience* 5, 195–206.

Stapp, H. P. (1993). A quantum theory of the mind-brain interface, in *Mind, Matter, and Quantum Mechanics*. Springer-Verlag, Berlin, pp. 145-172.

Stern, D. (1985). *The Interpersonal World Of The Infant.* New York: Basic Books.

Stern, D. N. (2004). The *Present Moment In Psychotherapy And Everyday Life*. New York: W.W. Norton.

Stern, K. & McClintock, M. K. (1998). Regulation of ovulation by human pheromones. *Nature* 392, 177-179.

Stewart, J. & Varela, F. J. (1991). Morphogenesis in shape-space. Elementary meta-dynamics in a model of the immune network. *J. Theor. Biol.* 153, 477-498.

Straker, L. M., Pollock, C. M., Zubrick, S. R. & Kurinczuk, J. J. (2006). The association between information and communication technology exposure and physical activity, musculoskeletal and visual symptoms and socio-economic status in 5-year-olds, *Child Care Health Dev.* 32(3): 343-51.

Strogatz, S. H. (1994). *Nonlinear Dynamics And Chaos: With Applications To Physics, Biology, Chemistry, And Engineering*. Reading, MA: Addison-Wesley.

Strogatz, S. H. (2003). *Sync. The Emerging Science of Spontaneous Order*. New York: Hyperion.

Tegmark, M. (2000). Importance of quantum coherence in brain processes, *Physical Reviews E* 61, pp. 4194-4206.

Thompson, E. & Varela, F. J. (2001). Radical embodiment: neural dynamics and consciousness, *Trends Cogn. Sci.* 5, 418-425.

Thompson, W. I. (1996). *Coming Into Being: Artifacts And Texts In The Evolution Of Consciousness*. New York: St. Martin's Press.

Tomasello, M. (2003). *A Usage-Based Theory of Language*. Cambridge, MA: Harvard University Press.

Tsuda, I. (2001). Toward an interpretation of dynamic neural activity in terms of chaotic dynamical systems, *Behavioral Brain Sciences* 24 (5): 793-810.

Turing, A. M. (1996). *The legacy of Alan Turing*. Oxford: Clarendon Press.

Umezawa, H. & Vitiello, G. (1986). *Quantum Mechanics.* Napoli: Bibliopolis.

van Gelder, T. (1998). The dynamical hypothesis in cognitive science. *Behavioral and Brain Sciences.*

Varela, F. J., Thompson, E. & Rosch, E. (1991). *The Embodied Mind, Cognitive Science And Human Experience.* Cambridge, Mass: MIT Press.

Varela, F. J. (1999). Cognition without representations, *Riv. Biol.* 92, 511-512.

Varela, F. J., Lachaux, J. P., Rodriguez, E. & Martinerie, J. (2001). The brainweb: phase synchronization and large-scale integration, *Nat.Rev.Neurosci.* 2 (4): 229-239.

Verhulst, F. (1994). Metaphors for psychoanalysis, *Nonlinear Science Today* 4, 1-6.

Vitiello, G. (1995). Dissipation and memory capacity in the quantum brain model, *Internat. J. Modern Phys. B* 9, 973-989.

Vitiello, G. (2001). *My Double Unveiled: The Dissipative Quantum Model of the Brain.* Amsterdam: John Benjamins.

Vitiello, G. (2002). *Dissipative Quantum Brain Dynamics, No Matter, Never Mind* (K. Yasue, M. Jibu and T. Della Senta, eds.), Amsterdam: John Benjamins, pp. 43-61.

von Neumann, J. (2004). *Mathematical Foundations of Quantum Mechanics.* Princeton University Press, 11:3829–34.

Ward, L. M. (2002). *Dynamical Cognitive Science.* Cambridge, Mass: MIT Press.

Watts, D.J., Strogatz, S. H. (1998) Collective dynamics of 'small-world' networks. *Nature,* Jun 4;393(6684):409-10.

Webber, C. L., Jr. & Zbilut, J. P. (1994). Dynamical assessment of physiological systems and states using recurrence plot strategies, *Journal of Applied Physiology* 76, 965-973.

Welsh, D. K. (1995). Individual neurons dissociated from rat suprachiasmatic nucleus express independently phased circadian firing rhythms, *Neuron.,* Vol. 14, 697-706.

West, B. J. (2006). *Where Medicine Went Wrong, Rediscovering The Path To Complexity.* New Jersey: World Scientific.

Whitehead, A. N. (1978). *Process and Reality.* New York: The Free Press.

Wiesendanger, M. (2006). Eccles' perspective of the forebrain, its role in skilled movement, and the mind-brain problem, *Prog. Neurobiol.* 78, 304-321.

Wills, C. L., Cacucci, F., Burgess, N. & O'Keefe, J. (2005). Attractor dynamics in the hippocampal representation of the local environment, *Science* 308 (5723): pp. 873-876.

Winnicott, D. W. (1958). *Collected Papers: Through Paediatrics To Psycho-Analysis.* London: Tavistock Publications.

Winnicott, D. W. (1971). *Playing and Reality.* London: Tavistock Publications.

Winslow, J. T., Hastings, N., Carter, C. S., Harbaugh, C. R. & Insel, T. R. (1993). A role for central vasopressin in pair bonding in monogamous prairie voles, *Nature* Oct. 7, 365(6446):545-8.

Wittgenstein, L. (1967). *Philosophical Investigations.* Oxford: Oxford University Press.

Yoo, H. J., Cho, S. C., Ha, J., Yune, S. K., Kim, S. J., Hwang, J., Chung, A., Sung, Y. H. & Lyoo, I. K. (2004). Attention deficit hyperactivity symptoms and internet addiction, *Psychiatry Clin. Neurosci.* 58(5):487-94.

Zanette, D. H. & Mikhailov, A. S. (1998). Condensation in globally coupled populations of chaotic dynamical systems, *Phys. Rev. E* 57(1), 276-281.

Zanette, D. H. & Mikhailov, A. S. (1998b). Mutual synchronization in ensembles of globally coupled neural networks, *Phys. Rev. E* 58(1), 872-875.

Zanette, D. H. & Morelli L. G. (2003). Synchronization of coupled extended dynamical systems, *International Journal of Bifurcation and Chaos*, Vol. 13, No. 4.

Zanette, D. H. & Mikhailov, A. S. (2004). Dynamical systems with time-dependent coupling: clustering and critical behaviour, *Physica D* 194, 203-218.

Zbilut, J. P., Sirabella, P., Giuliani, A., Manetti, C., Colosimo, A. & Webber, C. L., Jr. (2002). Review of nonlinear analysis of proteins through recurrence quantification. *Cell Biochem. Biophys.*,36, 67-87.

Zerubavel, E. (1981). *Hidden Rhythms, Schedules And Calendars In Social Life.* Chicago: University of Chicago Press.

Glossary

Aconscious

Contents and processes not repressed, not directly retrievable to consciousness, and embedded via psycho-physiological storage.

Affordance

All action possibilities which are latent in a specific environment, dependent or not on the individual ability to recognize them.

Assimilation

Cognitive and behavioral patterns are incorporated in implicit memory and motor schemes.

Attachment

The proximity to a person: attachment theory is concerned with the development of proximity relationships.

Attractor

A trajectory along which a dynamical system evolves after a long enough time.

Binding

Synchronous activity of neurons and neuronal ensembles.

Bipersonal

Shared events transcending individual members in a couple.

Cascading

A consecutive series of events or reactions.

Chaos

Complex causality of irregular deterministic behavior, highly sensitive to initial conditions.

Consciousness

The perception of a relationship between self and others or self-awareness.

Coupling

A process occurring when two systems repeatedly interact in a non-destructive way over a period of time.

Dissipative

A dynamical system where waves or oscillations lose energy over time, due to the action of friction or turbulence.

Dissolution

Biological and cognitive techniques for inducing suspension and change in attachment patterns.

Embodiment

A position in cognitive science stating that the mind emerges out of the interplay between brain, body and world.

Enaction

Knowledge that comes through action and is constructed on motor skills, in interaction with the environment.

Entrainment

The process whereby two interacting systems start coevolving.

Field

A set of elements and/or a region of space-time characterized by the same property.

Hyperstructure

A set of interdependent structures.

Meme

A unit of cultural information, practice or idea, that is transmitted verbally or by repeated action from one mind to another.

Mesoscopic

The mesoscopic scale lies between the macroscopic scale of the world we live in, and the atomic scale.

Mind Force

Influence resulting from the massive synergy of neurocognitive interactions.

Mindfulness

Full awareness of one's thoughts, actions or motivations in the present moment.

Network

A complex, interconnected or interacting system.

Neuroplasticity

Changes occurring in the organization of the brain as a result of experience.

Oscillator

Producing repeated and eventually periodic behavior.

Percolation

To spread slowly or gradually.

Quantum

A quantity considered as a discrete unit.

Self

The person, its peculiar qualities or the consciousness of identity.

Semantics

The study of meaning.

Semiosis

Any form of activity, conduct, or process that involves signs.

Synchronization

Coordination of events in space and time.

Wave

Event that propagates through space and time, usually with transference of energy.

Index